WESTERN GEOPHYSICAL

DEVELOPMENTS IN
GEOPHYSICAL EXPLORATION METHODS—5

CONTENTS OF VOLUMES 2 TO 4

Volume 2

1. Determination of Static Corrections. A. W. ROGERS
2. Vibroseis Processing. P. KIRK
3. The l_1 Norm in Seismic Data Processing. H. L. TAYLOR
4. Predictive Deconvolution. E. A. ROBINSON
5. Exploration for Geothermal Energy. G. V. KELLER
6. Migration. P. HOOD

Index

Volume 3

1. Underground Geophysics of Coal Seams. A. K. BOOER
2. Interrelationship of Resistivity and Velocity Logs. A. J. RUDMAN
3. Focused Resistivity Logs. A. ROY
4. Gamma-Ray Logging and Interpretation. P. G. KILLEEN
5. Acoustic Logging: The Complete Waveform and Its Interpretation. D. RADER
6. Electrical Anisotropy: Its Effect on Well Logs. J. H. MORAN and S. GIANZERO
7. Borehole Geophysics in Geothermal Exploration. W. S. KEYS
8. Measurement and Analysis of Gravity in Boreholes. J. R. HEARST and R. C. CARLSON

Index

Volume 4

1. Estimation of Sulphide Content of a Potential Orebody from Surface Observations and its Role in Optimising Exploration Programmes. E. GAUCHER
2. A Sweep-Frequency Electromagnetic Exploration Method. I. J. WON
3. The Magnetic Induced Polarisation Method. H. O. SEIGEL and A. W. HOWLAND-ROSE
4. Broadband Electromagnetic Methods. J. W. MOTTER
5. Radon Mapping in the Search for Uranium. W. M. TELFORD

Index

DEVELOPMENTS IN GEOPHYSICAL EXPLORATION METHODS—5

Edited by

A. A. FITCH

Consultant, Formerly of Seismograph Service (England) Limited, Keston, Kent, UK

APPLIED SCIENCE PUBLISHERS
LONDON and NEW YORK

APPLIED SCIENCE PUBLISHERS LTD
Ripple Road, Barking, Essex, England

Sole Distributor in the USA and Canada
ELSEVIER SCIENCE PUBLISHING CO., INC.
52 Vanderbilt Avenue, New York, NY 10017, USA

British Library Cataloguing in Publication Data

Developments in geophysical exploration methods.—
(Development series)
5
1. Prospecting—Geophysical methods
I. Fitch, A. A.
622'.15   TN269

ISBN 0-85334-216-4

WITH 160 ILLUSTRATIONS

© APPLIED SCIENCE PUBLISHERS LTD 1983

The selection and presentation of material and the opinions expressed in this publication are the sole responsibility of the authors concerned.

All rights reserved. No part of this publication may be reproduced, stored in a retrieval system, or transmitted in any form or by any means, electronic, mechanical, photocopying, recording, or otherwise, without the prior written permission of the copyright owner, Applied Science Publishers Ltd, Ripple Road, Barking, Essex, England

Printed in Great Britain by Galliard (Printers) Ltd, Great Yarmouth

PREFACE

The first phase of geophysical exploration is carried out on the surface. The subsurface in this phase is known only by geological inference, since it is inaccessible to direct observation. D. W. Strangway presents an interesting and effective method: audiofrequency magnetotelluric (AMT) sounding. This is a surface exploration method making use of a natural energy source derived ultimately from thunderstorms.

In the second phase of geophysical exploration, when some work in the subsurface has begun, it is often advantageous to make geophysical observations underground. By these means the observations can be made closer to the objective; often quite specialised methods must be developed. To this phase belong the chapters by D. J. Buchanan, who writes of fault detection in coal seams, and A. Hussain, who writes of underground gravity surveys. A quite different contribution to this phase of exploration is by J. G. Conaway on the improvement of geophysical logs by filtering.

T. E. Owen writes of a troublesome problem which can appear in either phase of exploration—discovery and exploration of caves and abandoned mine workings—in which a wide range of surface and subsurface methods are employed.

The editor takes this opportunity to thank the busy geophysicists who have spent time in preparing their contributions to this volume.

A. A. FITCH

CONTENTS

Preface v

List of Contributors viii

1. In-seam Seismology: A Method for Detecting Faults in Coal Seams 1
   D. J. BUCHANAN

2. Underground Gravity Surveys 35
   A. HUSSAIN

3. Digital Filtering of Geophysical Logs 65
   J. G. CONAWAY

4. Audiofrequency Magnetotelluric (AMT) Sounding . . . 107
   D. W. STRANGWAY

5. Detection and Mapping of Tunnels and Caves . . . 161
   T. E. OWEN

Index 259

LIST OF CONTRIBUTORS

D. J. BUCHANAN
> Head of Geophysics Group, Mining Research and Development Establishment, Ashby Road, Burton-on-Trent, Staffordshire DE15 0QD, UK

J. G. CONAWAY
> Section Manager, Well Logging Research, Seismograph Service Corporation, PO Box 1590, Tulsa, Oklahoma 74102, USA

A. HUSSAIN
> Digital Exploration Ltd, Digicon Building, Portland Road, East Grinstead, Sussex RH19 4HG, UK

T. E. OWEN
> Director, Department of Geophysics, Southwest Research Institute, 6220 Culebra Road, San Antonio, Texas 78284, USA

D. W. STRANGWAY
> Professor of Geology and Physics, University of Toronto, Toronto, Ontario, Canada M5S 1A1

Chapter 1

IN-SEAM SEISMOLOGY: A METHOD FOR DETECTING FAULTS IN COAL STEAMS

D. J. BUCHANAN

National Coal Board, Mining Research and Development Establishment, Burton-on-Trent, Staffs, UK

SUMMARY

Modern longwall coal mining is capital intensive and demands regions of coal in which the seam is not disturbed by faulting. Unfortunately, faulting occurs frequently, but if it can be mapped prior to mining, then its effects can be minimised by taking it into account during mine planning. The geophysical problem posed is therefore succinct: develop a method or methods, capable of mapping coal-seam discontinuities. In-seam seismics is such a method. Coal is a low velocity, low density medium relative to the surrounding country rock. Thus it acts as a waveguide to seismic waves generated by a source located in the seam. Hence guided or 'channel' waves propagate, and are reflected or scattered by acoustic discontinuities that may be present in the seam. The reflected signals are recorded and after computer processing reveal details of the plan position of faulting within the seam. This method was originally suggested over two decades ago, but only recently has it been developed to the stage where it can be used as a matter of routine. This chapter contains a description of its implementation with particular reference to the UK.

1. INTRODUCTION

1.1. Deep Coal Mining

A decade ago the economics of energy supply were based on cheap oil and a slow increase in the role of nuclear power. The 'contrived oil crisis' (see

Ezra[1]) resulted in the end of cheap oil. A growing public doubt, orchestrated by a powerful antinuclear lobby together with technical difficulties, has meant a slower rate of growth in the nuclear industry than that anticipated. Consequently the coal industry has been transformed from one steeply in decline to one in ascendency with a vital part to play in the World economy. Within the restricted field of exploration geophysics, this is recognised by the increasing number of papers that are being published or presented on coal geophysics.

The problem with coal is not finding it but mining it. As Ziolkowski[2] has recognised, coal geophysics is centred mainly on revealing geological structure in order to aid the mining operation. To appreciate this point a knowledge of mining methods is required and good description of this has been provided by Ziolkowski in Vol. 1 of this series. A brief recapitulation will suffice here.

If the seam is shallow, the overburden can be removed and the coal extracted. This is an extremely profitable method and is popular in the wide open spaces of the USA and Australia. In the UK only a few seams are shallow, and the population density is high. Consequently, opencast or strip mining accounts for only about 12% of production (1980/81).

If the coal is deeper, access to it is created by means of a shaft or a sloping tunnel. In 'bord and pillar' mining a series of tunnels is then driven in-seam leaving pillars about 30 m by 30 m to support the roof. At a later stage some of these pillars are removed. Again this method is widely used in the USA and Australia, but it is hardly practised at all in the UK.

In fact about 80% of UK coal comes from highly mechanised longwall coal faces. These consist of two parallel roadways about 200 m apart joined by a third road, along which a coal cutter travels back and forth carving coal which is then removed by a conveyor belt system. The area just mined is supported by massive hydraulic jacks. As mining progresses these are moved forward to allow the overburden behind it to subside slowly. This method is highly capital intensive, for a face can cost several million pounds sterling to develop and equip.

These three methods are by no means the only ones available, but they are certainly the most popular. This is not the right place to consider the advantages and disadvantages of each method, but we may note in passing that longwalling is widely practised in Europe as a whole because of the depths at which the coal seams occur. There is evidence that in other parts of the world, the mining fraternity is turning from bord and pillar to longwalling.

There are two principal forms of longwalling, although other variations are possible. On 'advancing' faces the parallel tunnels and the face advance at the same rate, with the ends of the tunnels only a few metres ahead of the face. On 'retreat' faces the parallel tunnels are driven first and then joined by the face, which then works back into the area encompassed by the tunnels. Figures 1 and 5 of ref. 2 illustrate this.

Longwall faces achieve a high percentage extraction in the region of the face but they cannot readily accommodate variations in the seam geometry. For example, an undulating seam is difficult to mine, as is one in which variations of seam thickness occur. But by far the most serious geometrical problem is that of geological faulting. A throw equal to only the seam thickness means that the face encounters a stone wall. The loss in production whilst the face is being re-located at the correct horizon is enormous. Thus faulting of apparently negligible proportions has a vital effect on coal mining economics.

Faults are not the only features that cause problems. Sand channel washouts, dykes, swilleys, sills, seam splits and rolls all affect mining adversely, but in the UK faults are the major source of worry. The average seam thickness in the UK is a little over one metre. In such a seam, a fault with a throw of up to the seam thickness can sometimes be mined through with a consequent loss in production. A larger throw usually leads to the abandonment of the face. In thicker seams, faults with throws less than seam thickness may lead to closure, while in thin seams faults up to three times the seam thickness may be carried.

Estimates of the economic effect of faulting vary and are subject to considerable controversy within the mining industry. However, all are agreed that a substantial increase in profitability would result if faults could be mapped in advance. Faces would then be laid out to avoid faulting and so reduce risk capital. This is an ambitious goal and one that will not be realised by geophysics alone. A combination of geophysics, geology and mining must be used if the aim is to be achieved. Progress is fraught with technical difficulties and human obstacles. Mining engineers are by nature conservative and do not readily respond to the structure estimates pushed at them by geologists and geophysicists. Fortunately, thanks to some far-sighted top-level mining engineers, coal mining geophysics has made substantial progress.

In this chapter we examine just one aspect of coal mining geophysics, viz. the in-seam seismic (ISS) method, but before doing this it will be useful to list some of the other methods for finding faults.

1.2. Exploration Methods

The surest way to prove the ground is to mine it, but this begs the question. Mineable coal, however, nearly always occurs in several different seams at various horizons. In old collieries several seams may be worked out and the fault pattern for those seams known. Fault hade angles can be measured and the fault patterns projected into the seam of interest. This method is widely practised by mine geologists, but statistics show that it is not always reliable. Geological plans indicate that not all faults occur at all horizons.

A second mining method of exploration is to use retreat faces. The parallel tunnels probe the ground prior to the face being established. Abortive face development can be avoided if faults are revealed by these roadways. There are however four disadvantages to retreat mining. Firstly, production is small whilst the roads are being driven. Secondly, a retreat face may sterilise the ground beyond it, since roadway drivage through a mined area is difficult. Thirdly, severe roadway stability problems may occur. Lastly, there are plenty of examples to show that Murphy's Law operates well in these circumstances, and faults can run sub-parallel to the main roadways. These may not be revealed until the face is developed and underway.

There are several ways in which surface boreholes can be used. Some structure can be deduced by correlating logs from different holes, but the fine structure will not be revealed. Boreholes are expensive (in the UK, £50 000–£100 000 each), and therefore the ground cannot be peppered with them. By using several holes it is possible to build up, over an extended period of time, a seismic array of geophones located *within* the relevant coal seam. Shots may then be fired from *within* the coal seam from an adjacent borehole, or from a nearby face if there is one. Thus the ISS method may be practised from surface boreholes. Experiments along these lines using both seismic and electromagnetic sources have been conducted by several research groups.[3-6]

Underground long-range horizontal drilling within the seam has also been advocated. This procedure is difficult and in any case samples only a small volume. Nevertheless, advances in the technology of surveying and guiding have been made.[7] The method will be best used in conjunction with seismic predictions. Drilling can be aimed at the target revealed by seismic work, so verifying, one way or the other, the predicted structure. This will be particularly useful for ambiguous seismic results. Various short range probes may be inserted into boreholes. Suhler *et al.*[8] have developed radar, acoustic and resistivity probes. These effectively enlarge the diameter of the hole.

The use of *surface* geophysics to delineate sub-surface features is well known in the oil industry. Similar techniques are now being applied with great success in the coal industry. Because of the depth at which coal and oil-bearing strata occur, most methods are based on the reflection or refraction of seismic waves. There is, however, a little work on electromagnetic and magnetotelluric methods. Several authors[9-12] have reported on the application of reflection seismics, and this work has been summarised by Ziolkowski[2] in an earlier volume in this series. It is thus possible to detect throws of the order of several metres in coal seams if the seismic data available is of *high quality*; however, such data is not always obtainable and in any case is expensive, although the cost in loss of production if such data is not obtained is potentially even greater. Irregular surface weathering, multiple scattering and a bad shooting environment such as Bunter Sandstone all lead to complex or even bad data and impede the process of interpretation.

Consideration has been given to the application of geophysical techniques at the coalface itself. The range required for detection (a few hundred metres) eliminates most methods. In 1975 experiments[13] on in-seam radar were conducted in the UK. These showed attenuation levels of 10 dB/m at 100 MHz resulting in a detection range in the reflection mode of only a few metres. This has recently been improved considerably[14] by the application of signal processing techniques and better equipment, but the range is still small. The future use of radar lies in providing high resolution, short-range information. Finally then we have the in-seam, or channel wave seismic method.

1.3. The In-Seam Seismic Method

In a short but classic paper[15] published in 1955 Evison reported that he had generated and recorded seismic waves in a New Zealand coal seam. Analysis of the wavetrain showed that it could be identified as a guided Love wave. Evison made the prescient observation that: 'Guided waves may find useful applications in mining'.

The principle of the ISS method is simple. Mine roadways can be used to build up extended arrays of sources and detectors in the coal seam itself. The possibility thus exists of turning surface seismic reflection methods through ninety degrees, to probe horizontally for faults within the coal seam. Even faults with a small throw will present a large horizontal acoustic impedance and so reflect energy back along the coal seam to the geophone array.

In 1963 Krey published his account[16] of the theory of mode propagation of seismic waves in coal seams. He showed that for the geometry of a coal seam surrounded by rock, there exists a set of waves whose quintessential feature is that they are confined to the coal seam and its surrounds. These waves are variously called 'channel' or 'seam' waves and they are dispersive; i.e. the velocity is a function of frequency and hence a channel wave changes shape as it propagates.

A variety of factors conspired to prevent rapid progress in the subject. Perhaps the chief reason was that the coal industry was in decline. There was insufficient motivation to develop a routine service for the industry. Progress was also impeded by a failure to recognise that the technique was substantially different to the seismic practice in use at that time. There are three main differences: the survey *objective*, the survey *method* and the *nature* of channel waves.

The survey objective is to image structures relatively near to the survey line, i.e. a few hundred metres. This is in contrast to oil exploration seismology (which has been the instigator of almost all seismic exploration developments) where the object is to map oil bearing structures deep within the earth's crust, i.e. a few thousand metres. Furthermore, in channel wave seismology there is no *a priori* reason to believe that the target is almost parallel to the survey line. The underground environment is harsh. As Fig. 3 of ref. 2 shows, space and access are severely limited. This necessarily influences data acquisition. Very stringent mine safety regulations apply, requiring specialised equipment developments. Finally, specialised data processing methods are necessary both to cope with dispersive propagation and the survey objectives.

A number of groups throughout the World have been engaged on research into the ISS method. This work has been concentrated in Europe where longwalling preponderates, and the need to solve the problem is greatest. An extensive series of references is given by Mason *et al.*[17] Rather than provide an historical review, the rest of this chapter is devoted to the ISS method in its current form (early 1982).

2. CHANNEL WAVES

Coal is a soft, low density material which is found in seams, sandwiched between harder and denser materials such as mudstone or sandstone. The measured seismic velocities in coal vary, but typical values are 1 km/s and 2 km/s for S and P waves respectively. The surrounding country rock may have velocities double these values. The acoustic impedance contrast is typically three or four to one.

If a point source of body waves is located in the coal seam, then the waves emitted will be partially reflected at the upper and lower boundaries of the seam. Energy will be propagated down the seam as the wave bounces back and forth. For initial ray paths that are near vertical, the reflection is subcritical and some energy is transmitted through the boundary as the wave propagates down the seam. These reflections form the so-called 'leaky modes'. For more horizontal initial ray paths, the critical angle is exceeded and total internal reflection occurs. These waves are termed the 'normal modes'. A channel wave is normally understood to be a normal mode. In addition critically refracted, or head, waves also occur.

There are several theoretical approaches to this problem. For example, Krey[16] has used normal model theory to derive the dispersion relation, whereas Buchanan[18] has adopted a Green's function approach. From a mathematical standpoint, it is the boundary conditions of continuity of stress and displacement that give rise to all the interesting phenomena.

For the simple case above, SH, or Love, channel waves exist with the plane of polarisation transverse to the direction of propagation. In addition, P–SV, or Rayleigh, channel waves occur with the plane of polarisation in the sagittal plane. Note that separate P and SV channel waves do not exist: the boundary conditions will not permit it.

The most fundamental property of channel waves is that they are dispersive. Consider a wave $\psi(x, t)$ as defined by its Fourier transform

$$\psi(x, t) = \int_{-\infty}^{\infty} dw A(x, w) \exp\left[i(kx - wt)\right] \quad (1)$$

A is the amplitude spectrum, k is the wavenumber and w is the frequency. The latter two variables are related by the equation

$$f(k, w) = 0 \quad (2)$$

In eqn (2) f is some function determined by the physics of the problem. For example, in channel wave theory f is determined by the boundary conditions, whilst in phonon propagation the crystal structure determines f. In one space dimension the equation represents a curve in the w–k plane. In N space dimensions it represents an N-dimensional surface embedded in the $(N + 1)$-dimensional w–$\mathbf{k}_N$ space. For simplicity we consider one space dimension.

The velocity $V = w/k$ is called the phase velocity and, as the name implies, it is the velocity with which the surfaces of constant phase advance. The group velocity $V_g = \partial w/\partial k$ plays a vital role in channel wave propagation. If f represents a straight line then $V = V_g$ and all higher derivatives of w are zero. Such a wave propagates with its shape unaltered: this is normal, non-dispersive propagation. (In this treatment we are ignoring geometrical spreading, and attenuation losses.) Curvature in f implies dispersion. For channel waves the importance of V_g lies in the fact that it is the velocity of energy propagation. (This is not a universal result for dispersive propagation as a whole. For example, in the propagation of electromagnetic waves in dielectrics V_g can sometimes be greater than the velocity of light. This is called anomalous dispersion, and V_g has no physical significance.)

It is not difficult to see that a hypothetical wave with V different from V_g but with higher derivatives of w equal to zero, propagates with a uniform shape. Therefore, in dispersive propagation, change of shape of the wave comes solely from the second and higher derivatives of w. This result has been used by Mason et al.[19] to compress channel waves by passing the signals through a digital filter which removes the second derivative. Dispersion is thus greatly reduced, for only the high order derivatives remain and these are relatively small.

Dispersion is apparently a serious problem to be overcome in implementing channel wave exploration. The waves become more and more diffuse as they propagate, thus making identification of arrival times difficult. Energy conservation implies signal amplitudes less than those that would be observed if no dispersion were present in a medium of the same dimensionality. High quality equipment is therefore required to capture the data. Multiple propagating modes can and do occur. These disadvantages are offset by the fact that in a coal seam the propagation is essentially two-dimensional rather than three-dimensional. If reflected channel waves can be identified then the interpretation problem is eased, for they can only have come from the coal seam and not from some structure above or below the seam of interest.

This section provides only the very briefest introduction to the theory of channel waves. The interested reader must consult the original publications for further details. Buchanan[18] has examined attenuated SH channel waves; Lagasse and Mason[20] and Lagasse[22] have considered the effect of the roadway in which the source is placed and found that the dispersive equivalent of ground roll occurs; Franssens et al.[21] have examined the link between leaky and normal modes; Hasbrouck and Hadsell,[3] and their students[23-25] have used numerical techniques to generate synthetic seismograms for different complex structures; finally Dresen and his students[26-29] have carried out a series of elegant model experiments, and cast light on a variety of topics including reflection coefficients.

3. DATA ACQUISITION

The coal face is typically 200 m long. For probing ahead of the face, the seismic line is restricted to this distance. Theoretically, the roads leading up to the face can be any length, but in practice they tend to be less than 1 km with 700 m being typical. The face itself is congested, although the coal seam is, of course, well exposed. The approach roads are not so congested, but often the seam is not exposed. Sometimes iron sheets line the road; packs of stone and wood are occasionally used for support purposes; and frequently the floor 'heaves' or rises up. These three factors make for difficult access to the seam adjacent to the roads. The consequence is that the geophone and source arrays must be kept simple. It is not practicable to think in terms of sub-arrays for each geophone station. Booer[30] has also discussed the theoretical reasons for choosing single geophone stations.

As a seismic source, small explosive charges (100 g) are convenient. Persons not acquainted with the mining industry are sometimes surprised that shot firing is permitted, but in fact there is a wealth of experience of its use because prior to mechanised cutting, coal was mined by shot firing. By a happy coincidence the permitted explosives are a good acoustic match to coal.

Before looking in detail at the field survey method, it is necessary to have an appreciation of mine safety regulations. Apart from the obvious hazard of being hundreds of feet underground in a dark, congested tunnel, the main danger is the possible ignition of explosive gas, and electrical equipment is a possible source of sparks. To avoid explosive ignition, electrical equipment must conform to certain standards. Either the circuitry is such that under fault conditions incendive sparks cannot occur (this is protection by intrinsic safety); or the equipment is housed in massive steel boxes so that any explosion is confined to the interior of the box and cannot propagate explosively outside (this is protection by flameproofing); or the equipment is housed in steel boxes which contain an inert gas atmosphere at a pressure slightly higher than atmospheric (this is protection by purging). Additionally, equipment cannot be constructed from a light alloy such as aluminium, for such material can react with ordinary rust to produce an incendive spark. In some instances it is possible to obtain a *temporary* relaxation of some of these rules provided gas levels are closely monitored.

Commercial seismic equipment breaks all the rules. It is constructed from a light alloy for ease of transport, and power consumption is considerable. Nevertheless, such equipment has proven reliability, and that is essential for conducting in-seam surveys. To use this equipment underground the West Germans have followed the purging method,[31] whilst the National Coal Board in the UK has obtained temporary relaxation with eventual protection by flameproofing. The equipment used in the UK comprises a 24-channel Sercel SN338HR with 0·5 ms sampling, whilst the Germans use a DFS V with the same configuration. Both systems are very bulky. In both cases rechargeable batteries are used and these must also be protected.

The astute reader will suggest using a telemetry system to ease the protection problem, with the bulk of the equipment located on the surface. Unfortunately the systems that are available currently have either a limited range or use frequent repeater amplifiers. In addition special cable is required. The distance from shaft to coal face can be as much as eight miles. A routine service cannot be based on a method requiring the laying of so much special cable, and in particular putting a cable down a shaft is not straightforward.

The long term solution probably lies in an intrinsically safe, and therefore small, system such as that being developed by Rüter and Schepers.[32] Alternatively if a fibre-optic seismic telemetry system is available, and *if* fibre optics ever become commonplace in mines for other communication purposes, then such a system could greatly ease the acquisition problem.

The regulations also impinge on possible developments of an alternative source to explosive charges. Such a source (if electrical) would have to be fully protected. The lead time for development and approval would be lengthy; a controlled, swept frequency source, for example, will not appear overnight.

Two main types of survey are undertaken: 'transmission' surveys which involve the propagation of channel waves across a block of coal of known geometry, and 'reflection' surveys which probe areas in advance of mining. Transmission surveys provide vital velocity information, and wherever possible are carried out in an area adjacent to the reflection survey. In both cases the shots and geophones are located in short holes drilled horizontally into the seam. The spacings are dependent on the size of the area to be investigated. In general, the aim is to achieve six- or twelve-fold coverage. The detectors comprise three 28 Hz geophones arranged orthogonally within one steel or brass unit. Any combination of components can be connected. Good coupling to the sides of the hole is achieved either by means of a mechanical clamping device, or pneumatically. Geophones can be located anywhere within the colliery and coupled via cables to the recording equipment sited in a low gas area. Further details on equipment are given in refs. 31, 33 and 34.

4. DISPERSION ESTIMATION

We turn now to the subject of data processing. To analyse the reflection data, velocity information is required. This is obtained by estimating the dispersion relation from transmission channel waves.

Dispersion estimation methods fall into three classes: calculation from elastic constants, extraction from recorded data and guesses. In principle, calculation is the best method. The Thomson–Haskell[35,36] method can handle complex, layered structures. Unfortunately this approach requires a knowledge of the elastic constants *in situ* and this information is rarely available.

In the second class, several different techniques exist depending on what data is available. If only a single trace is recorded then phase-spectrum unwrapping techniques give the phase velocity directly, whilst Hilbert transform theory can be applied to narrow bandpass filtered data to give the instantaneous amplitude and hence group velocity. If multirace data is available, recorded over some spatial array, then a double Fourier transform[17] in time and space gives the $w-k$ curve directly, or a slant stack[37] gives the phase velocity.

In the third class, a functional form for the dispersion relation is hypothesised, and the parameters are established by fitting to it the recorded data. For example, the dispersion relation can be written[30] as a Taylor series in frequency truncated after the quadratic term. It is the coefficient of this latter term that controls the rate of dispersion, and this coefficient can be estimated[19] by finding a filter which maximally compresses the seismic signal. The first two coefficients of the Taylor series are of course simply related to the phase and group velocities. The advantage of a quadratic function is that it can be treated analytically, but other three-parameter functions may well represent the dispersion relation more accurately. For instance, the [1:1] Padé approximant[57] has three parameters and an infinite number of derivatives. Such rational

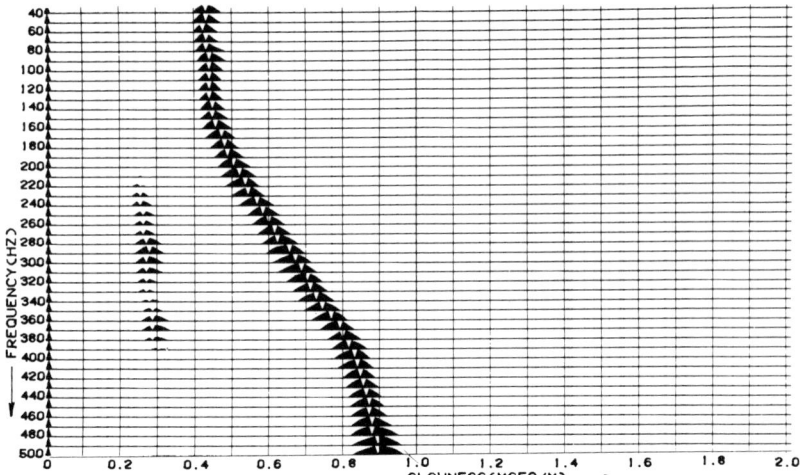

FIG. 1. Slowness stack of envelopes from transmission shear motion.

approximations are often better representatives of functions than simple polynomials. The technique now mainly used by the NCB falls into the second class. A transmission survey is carried out by firing separately about a dozen shots from different locations to an array of twelve geophone stations. Both horizontal components at each station are recorded. The first stage in processing this data (after noise removal and other pre-processing operations) is to rotate[38] each pair of traces to form two new traces as if recorded transverse to the shot–geophone line and along it. The transverse trace will contain predominantly shear arrivals. This trace is then analysed using Dziewonski's method:[39] Fourier transform, narrow bandpass filter, Hilbert transform and inverse Fourier transform. The instantaneous amplitude (envelope) is calculated, and the time variable is scaled to group slowness using the known shot–geophone distance. This produces a set of envelope-slowness traces, one for each bandpass filter. The calculation is repeated for all traces and shots. Envelope-slowness traces from the same filter are then stacked, and the result is finally displayed as a frequency-slowness stack as shown in Fig. 1.

This result is typical and shows a fast, non-dispersed arrival (the head wave), followed by the dispersed channel wave. The longitudinal component can also be analysed in a similar way giving the result shown in Fig. 2. In the latter instance the channel wave is usually less noticeable,

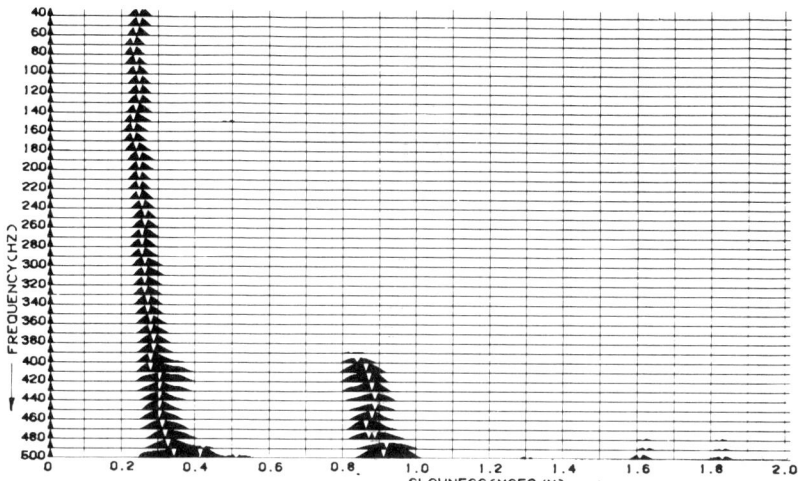

FIG. 2. Slowness stack of envelopes from transmission compressional motion.

showing that it is the SH variety. The head wave is normally stronger than in the transverse component, showing it to be a P wave. This result does not hold if the seam is anisotropic.[40] Thus the existence of anisotropy can be inferred from a lack of P and S separation in slowness stacks.

The detailed shape of the dispersion curve depends on the local physical properties and on the seam thickness. In a thick seam the curve is contracted towards the low frequency end. This has been observed in measurements made in Australian mines[41] and is in full accord with the theory. The part of the dispersion curve in which the velocity is changing most rapidly corresponds to the part with maximum dispersion. Sometimes this part of the curve is missing, presumably because the seismic energy has been dispersed in such a way that it cannot be discerned in the noise. The best reflection seismic results are obtained from the slightly dispersed, high frequency end of the spectrum. There can be considerable variations in the curves obtained from different sites, even those from within the same seam, showing the importance of local measurements.

There are several ways of determining the phase velocity, either directly or by derivation from the group velocity. The phase spectrum of a single trace is easily found modulo 2π, and if it can be unwrapped successfully, the phase velocity can be derived. The phase velocity can also be found by using seismograms recorded at two different places, although there are again ambiguities of 2π. As mentioned previously, slant stacks and double

Fourier transform methods are also applicable. The radial lag sum (RLS) imaging method has been described by Booer.[30] This is a technique for reconstructing images of sources (real or virtual) from a set of seismograms. The image that is produced is a function of the group-to-phase velocity ratio. Given that in a transmission survey the location of the source is known, the method can be used to find the velocity ratio by optimising the image. Finally, Buchanan and Jackson[42] have used an iterative technique to fit a sinusoid to the measured phase spectrum. This technique is well suited to a set of seismograms rather than a single trace.

In the Taylor expansion of the dispersion relation, the third term is called the chirp rate and is primarily responsible for the rapidity with which the signal becomes diffuse. Estimation of this term by recompression procedures has been discussed extensively.[17,19,30,43] A related approach, which is implemented entirely in the frequency domain, has also been recently introduced.[42] In this case dispersion is linked directly to curvature of the $w-k$ relation and compression of seismic signals is achieved by finding the transformation which minimises curvature. These two methods are independent of the travel-path length and can be applied to signals consisting of two arrivals (direct and reflection) with the same dispersion characteristic. A different technique has been adopted by Marschall and Schott.[44] They find the time domain operator equivalent to dispersion, and achieve compression by reversing the sign of the group delay in this operator and applying it to the recorded data. This operator is, however, dependent on the path length.

In the early history of the ISS method considerable effort was spent on developing processing methods to take account of dispersion. In particular, the compression of both transmission and reflection seismograms was studied. Theoretically, optimisation of one parameter (the negative chirp rate) ought to be sufficient for compression, but in practice it was found that considerable scatter occurred and that different parameter values compressed different traces optimally. Consequently, skill was necessary to process the data, and a semi-automatic method for processing several hundred *surveys* per year seemed unobtainable. However, experience showed that faults either reflected energy exceedingly well over a wide bandwidth in which case they were detectable without resort to sophisticated compression procedures, or else they reflected over a narrow bandwidth in which case the signal was scarcely dispersed. Good results, therefore, can be obtained by treating data pre-processed by a bandpass filter.

In other words, it turns out that in a large number of surveys dispersion can be treated by the simple expedient of measuring the group velocity and applying a narrow band filter prior to reflection analysis. Currently the role of compression lies in the fine tuning of results; a role which it shares with, for example, the recognition of anisotropic propagation. Ultimately, however, dispersion is one of the factors that control the range of the technique. For long range work results which are significantly better will be achieved if the signals are compressed. The same comments apply to the resolution obtainable.

5. PROCESSING REFLECTION SURVEYS

In the UK two main approaches to the processing of reflection surveys have been adopted. One is a migration procedure based on the Kirchhoff–Huygens approach to diffraction theory, and the other is a modification of CDP stacking.

A linear seismic array can be regarded as a diffraction grating. The recorded seismograms are records of the incoming waves to the grating. The object of processing this data is to produce the diffraction pattern which then needs to be interpreted. The difference between a transmission and reflection survey is that in the former the source of the waves is ideally a point, whilst in the latter it is extended. Of course data from reflection surveys also contains more noise.

In deriving the imaging equation from diffraction theory, three factors different from the usual monochromatic case must be taken into account: time dependence, two dimensional propagation and dispersion. The result for a finite array of geophones is (see the Appendix of ref. 45)

$$I(x,y) = \left| \sum_{n=1}^{N} U_n(t_n) \exp\left[iw_c(1 - C_g/C_p)t_n\right] \right|^2 \quad (3)$$

In this equation x and y are the coordinates of the point at which the image I is to be calculated, $U_n(t_n)$ is the signal at geophone n, C_g and C_p are the group and phase velocities at a central frequency w_c, and t_n is the travel time to the image point.

This equation takes the form of a 'lag-sum', and may be implemented in a variety of forms according to how the travel time t_n is calculated. In the case of data from a single seismic source (shot) radial, elliptical and mode-conversion lag-sums have all been implemented.[17,19,30] These schemes may be used variously to map sources either real or virtual, to image faulting in true plan position, and to locate faults using transmission rather than reflection data. Examples of their usage are given in the references alluded to above. These techniques were vital in proving that channel waves could be used to detect faulting. However, they were far from optimum in implementation, and considerable skill and time was required to use them. Extensions were required to reach the goal of a scheme capable of being relatively easy to use with the potential to process surveys quickly. To this end, adaptive lag-sum (ALS) and dynamic trace gathering (DTG) were developed.[45]

As its name implies ALS is an extension to the basic lag-sum procedures. Data from successive shots is processed by summing the maps resulting from each shot. Maps can also be focussed to image faults lying at a particular angle to the survey line by applying a weighting function to the data. Map reliability is assessed by counting the number of contributions made to each map point (x, y). The resulting composite map is usually displayed as a contour diagram. Examples of the technique are given in ref. 45.

Millahn and co-workers[46,47] have produced an interesting extension to lag-sum processing. Data is recorded in the two horizontal components at each geophone station. The rectilinearity and angle functions are then found.[38] Rays are back-propagated in the direction from whence they came, and by using the measured velocity, the shot image is found. The corresponding reflection point is easily calculated given the shot and geophone locations. Trace amplitudes can then be assigned to the reflection point and a map built up. The overall map should be an improvement on basic lag-sum maps since polarisation of recorded data is taken into account.

CDP stacking is a universal practice in surface seismic exploration and is designed to highlight seismic reflectors by improving the signal-to-noise ratio. This improvement comes about because reflection points in the earth are sampled many times. Unfortunately if the reflector is not parallel to the survey line, the calculated reflection points are not identical to the true reflection points. It may be shown,[45] for traces gathered at the point midway between shot and geophone, that the error d between the true and calculated reflection points is:

$$d = x^2 \sin(2\theta)/2L \qquad (4)$$

where $2x$ is the shot–geophone distance, θ is the angle of dip of the reflector and L is the perpendicular distance from the midpoint to the reflector. In standard surface applications x/L and θ are both small and hence d is small.

For in-seam seismics both of these conditions may be violated. As an example, consider a surface case with $x = 500$ m, $L = 4000$ m and $\theta = 10°$, then $d/x = 2\cdot1\%$; but the ISS case with $x = 100$ m, $L = 100$ m and $\theta = 45°$ gives $d/x = 50\%$. In addition CDP stacking does not take into account the differing group and phase velocities of dispersed signals even after recompression. Consequently a modified CDP stack, called dynamic trace gathering (DTG), has been introduced.[45] In the routine processing of reflection surveys DTG has been found to be more useful than the lag-sum methods. The basic calculation procedure is as follows.

A hypothetical target reflector at a certain angle of interest is defined by the processor. This target is subdivided into a number of equal length segments. For each seismic trace the reflection point on the reflector is calculated and the trace is assigned to the line segment containing that reflection point. This is a dynamic assignment dependent on the target of interest. Those traces belonging to the same segment are stacked after applying a normal moveout correction. In this way only traces with similar reflection points are stacked. The error in the calculated reflection points is limited at most to the size of the segment, and since this can be chosen at will; the error can be made as small as is desired at the cost of computer time. In the absence of *a priori* information on possible fault angles, a range of angles must be scanned followed by a refined search if something of interest is revealed.

Two methods may be used to take account of dispersion. Firstly, the phase roll factor

$$\exp[iw_c(1 - C_g/C_p)t]$$

may be applied to each trace before moveout correction and stack, or secondly, trace envelopes may be stacked in which case all phase information is lost. Both methods require prior filtering by a bandpass filter centred on w_c, and the whole calculation is usually repeated for different values of the central frequency w_c. Normally if w_c is large there is insufficient phase coherency to justify using the phase correction; it is better to stack envelopes.

The scheme is easily programmed to process large volumes of data with little or no operator interaction. Each stack is displayed as a space–space section which must be interpreted.

5.1. Examples

In previous papers[34,48] several examples of surveys have been given. In

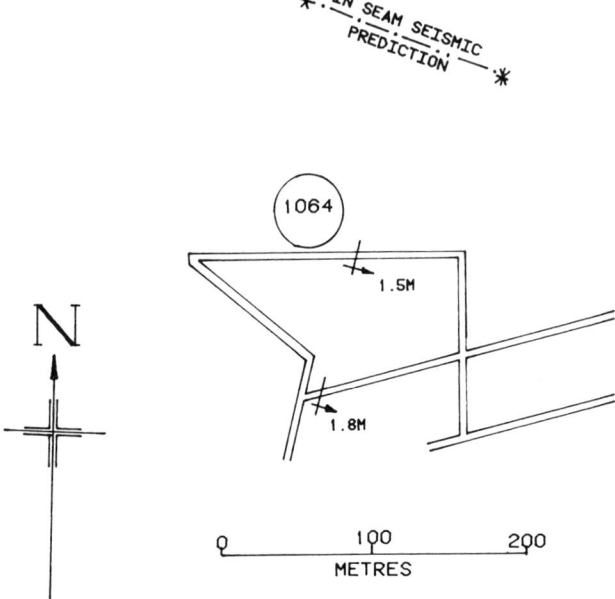

FIG. 3. Plan of site to be investigated: a reflection survey was carried out on the face AB.

this chapter three new cases are presented, each of which illustrates a different aspect of ISS.

The first example, Fig. 3, concerns a face AB which has been opened up but (at the time of the survey) had still to be equipped. The problem is to check the area to the north of the face for faulting. A reflection survey was carried out on the face. In this instance there was no suitable transmission site for velocity analysis, but at a previous survey in the same seam good velocity information had been obtained. The principal event that showed up in the DTG processed section is shown in Fig. 4. There are three features worthy of comment.

Firstly, the prominent event adjacent to the face. This is *not* indicative of faulting; it is simply the roadway mode, i.e. the underground equivalent of ground roll. The event bulges because in this case the roadway mode velocity is lower than the stacking velocity. This behaviour can be verified either by careful consideration of the stacking process, or else by processing synthetic data. The second noteworthy feature is the quality of the section, or lack of noise. This is not always the case, but because only close events are being imaged and there are no multiples to contend with, the quality of

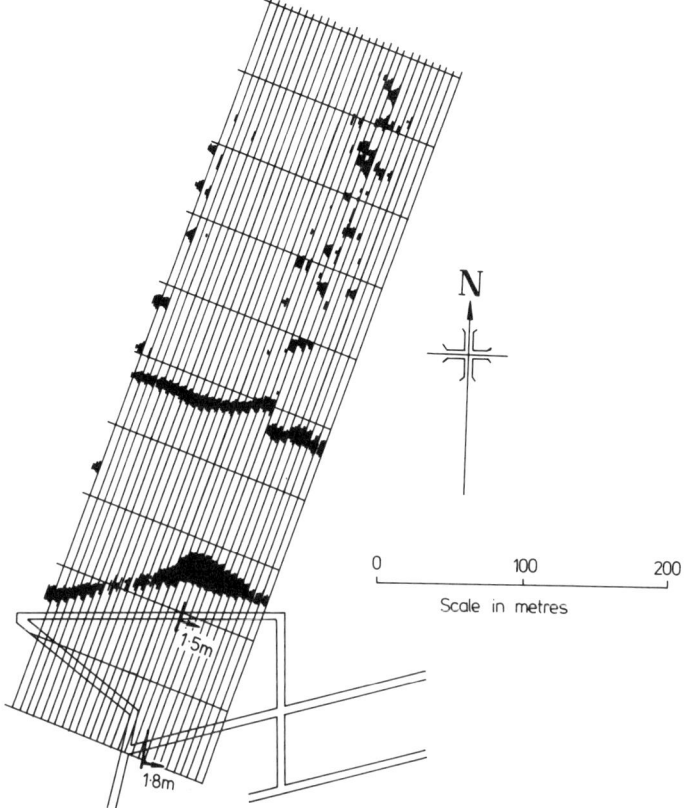

FIG. 4. DTG seismic section: note the appearance of the roadway mode as well as the reflection event.

the best sections far exceeds any surface section. The third feature is of course the broken event about 120 m from the east end of the face. This is interpreted as a large (face stopping) fault. The break in the event may be due to the intersection of this fault with an extension of the 1·5 m fault found on the face. There is also evidence of diffraction at the broken event, caused presumably by the fault intersection. Notice that the break does *not* lie on linear extrapolation of the 1·8–1·5 m fault. This fault could of course change direction, but the effect could also be the result of anisotropy: a point to which we will return later. The final interpretation is shown in Fig. 5.

Figure 5 is the diagram of main interest to colliery management. In

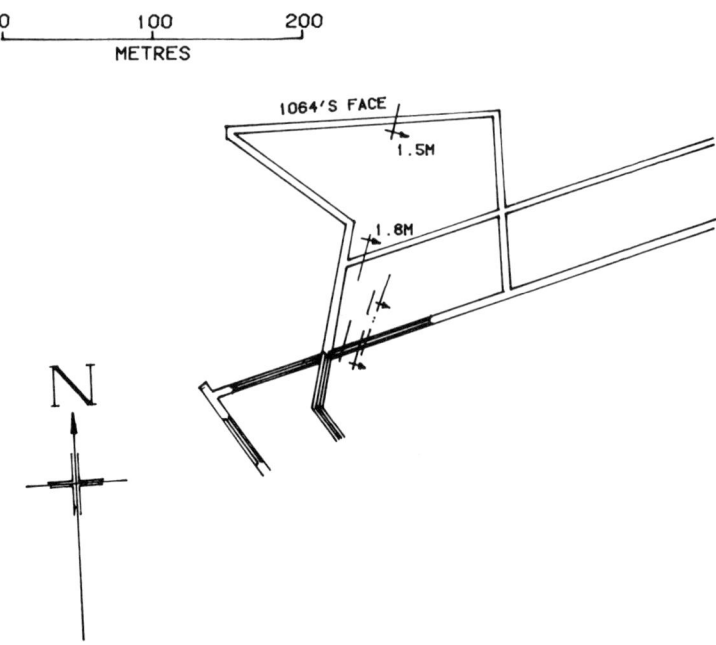

FIG. 5. Final interpretation: a fault is predicted with a high degree of confidence.

presenting the final result, it is vital to convey some feeling of confidence in the result. Consequently results are classified on a scale of 1 to 4 as follows:

1. —.—.—.   very confident
2. —..—..—..   confident
3. —...—...—... some doubt
4. .... considerable doubt

Confidence levels may change along the one seismic event. Assigning confidence levels is largely a matter of experience. As a rough guide reconsideration of mining plans should be contemplated for results in the first two categories but not for those in the latter two; however, mine management must have the final say. It may be that a result with the lowest confidence level has such a potential impact on plans, that action of some sort is warranted. For example, it may be necessary to verify the result by driving a roadway.

It is also vitally important to impart to colliery management an appreciation of the detection capabilities of ISS. Amongst other things, these depend on the structure and geometry of the target. The principal factors are:

Throw—unless the fault has a large enough throw it will not be detectable.

Hade angle—if the hade angle is large then the fault is effectively spread out in range from the survey line. The reflection may not be clear and there will be considerable ambiguity in position. It is the combination of hade and throw that is important. A (relatively) large throw coupled with a very large hade will be a difficult fault to detect.

Trend angle—the angle which the trend of the fault direction makes with the survey line must not be too large.

Fine structure—a fault that is rough in either section or plan will be frequency selective. If the optimum reflected frequency coincides with one that is difficult to measure (e.g. 50 Hz), then detection will be difficult.

Other features, in principle, other discontinuities can be detected, but

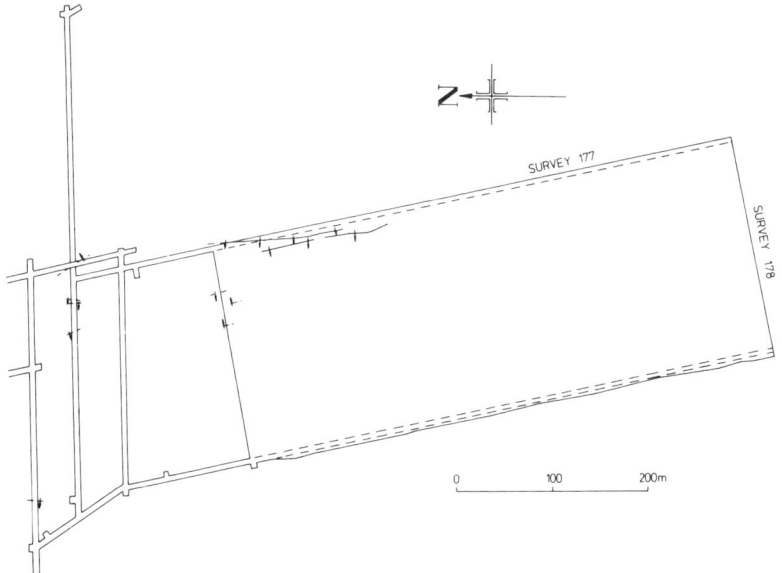

FIG. 6. Plan of site to be investigated: surveys were conducted to look to the side of the face (177) and ahead of the face (178).

currently ISS cannot differentiate between these events. Thus an event may be interpreted as a fault when it is in fact a dyke or a washout.

The second example, Fig. 6, illustrates a common use of the method. A survey (178) was undertaken to investigate the area ahead of the face to the south, and another (177) to investigate the area to the west where the next panel of coal is planned. A transmission survey was carried out over the block of coal near the start of the panel. The best section from survey 177 is shown in Fig. 7, and the final interpretation of both surveys is shown in Fig. 8. This indicates a possible fault about 295 m from the face but the confidence level is not high. Much more important is the confident prediction of faulting running almost parallel to survey line 177 and at a distance of 190 m. This is an extremely important result, for it is this fault that will determine the optimum width of the next panel to be mined.

The last example, Fig. 9, is also from a roadway AB leading up to a working face. As a result of mining in a different horizon, a fault close to AB

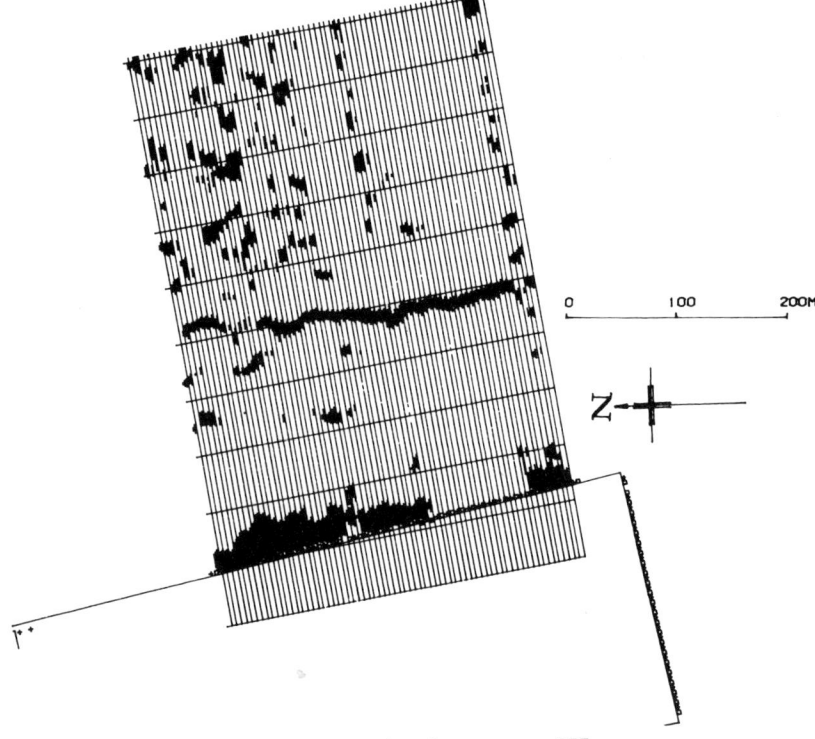

FIG. 7. DTG section from survey 177.

IN-SEAM SEISMOLOGY

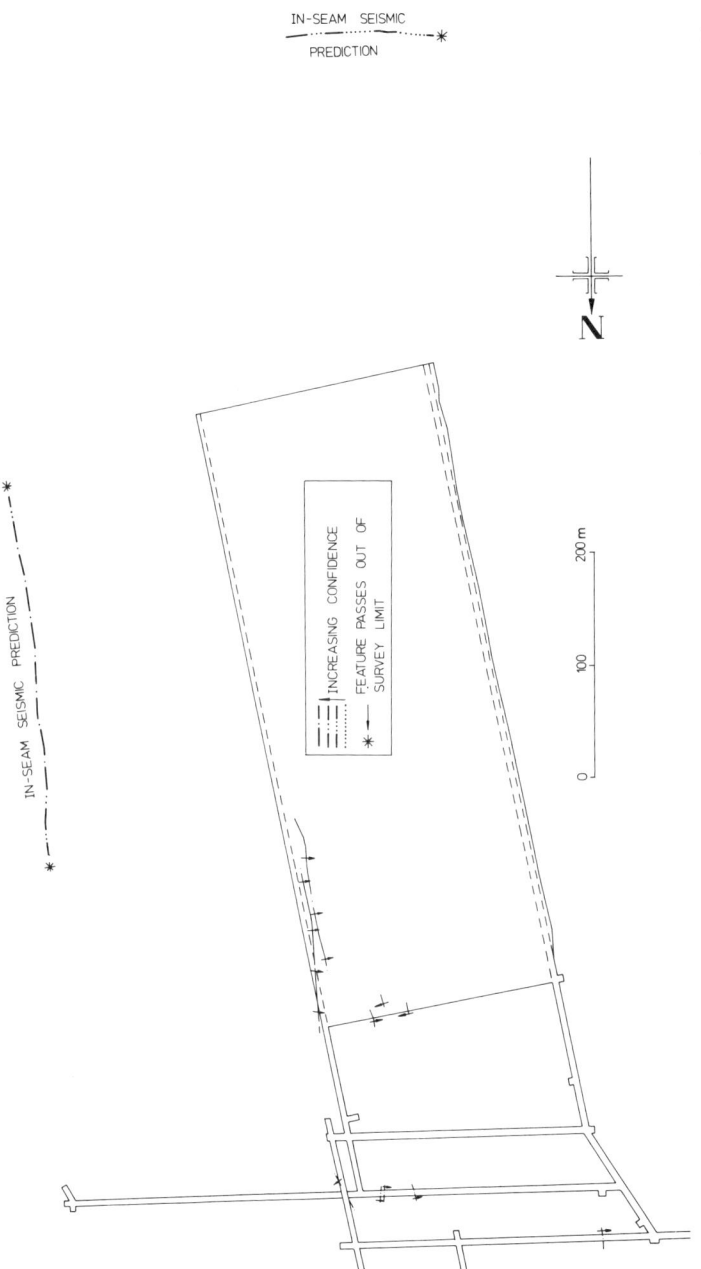

FIG. 8. Final interpretation of surveys 177 and 178: the confident prediction of a fault almost parallel to survey line 177 enables the width of the next panel of coal to be planned optimally.

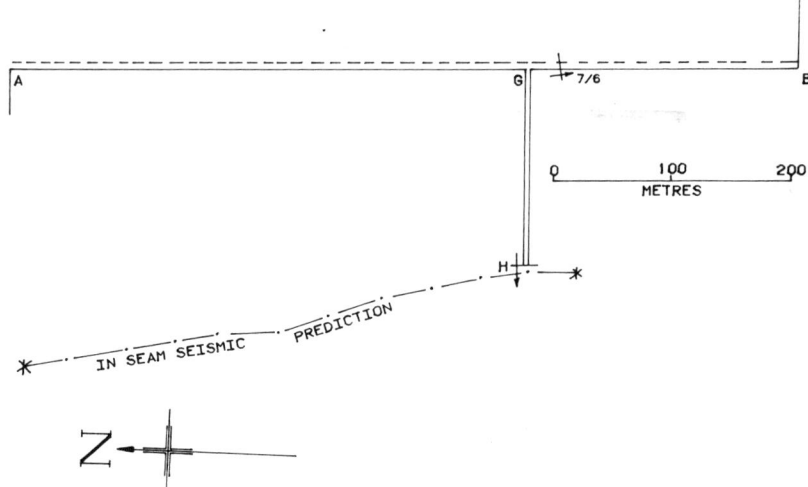

FIG. 9. A survey from AB produced the predicted fault as shown, subsequently verified by the roadway GH.

was expected. The expected position was such that mining of the coal adjacent to AB would not be economic. However, a seismic survey carried out from AB indicated faulting in the position shown, and roadway GH driven subsequently has verified this prediction to within 3 m. The predicted and verified fault position is such that a face between AB and the fault is an economic proposition. ISS has therefore provided a panel of coal that was previously written off as uneconomic.

One of the most satisfying features of ISS is that in a substantial proportion of cases, the correctness or otherwise of fault predictions can be verified directly by mining. Thus it is possible to establish meaningful statistics on the success rate of ISS. These surveys that have been checked by independent means have been awarded points reflecting the confidence weighting of the original seismic prediction. Hence confidence level 1 results are awarded four points (in either the success or fail column), level 2 results are three points, etc. By this estimate ISS has a success rate of almost 80%.

6. ADDITIONAL TOPICS

ISS has reached the stage where it can be, and is being, routinely exploited as a means to aid mine planning. There is of course scope for improvement, and in common with other branches of seismology, there are plenty of other research problems to be tackled. In this section we will outline some of the problems in which progress has been made.

6.1. Attenuation

The useful range of the method in a particular locality depends upon the rate at which channel waves are attenuated during passage through coal. A major source of worry in interpretation is the case when *no* faulting is detected. It is important to know at what range faulting would have been detected had it been present. Obviously a number of factors determine this distance, a major one being the attenuation or absorption rate.

Buchanan et al.[40] have measured attenuation in UK coal seams. Transmission data is required over a range of propagation distances. Each seismic trace is time-gated with a cosine bell in order to include only data from within a specified range of velocities centred on the channel wave velocity of interest. This data is then Fourier transformed. Data at a given frequency is plotted in the form of log amplitude against the product of frequency and range. A correction for cylindrical spreading is included. The frequency-range variable is chosen because an attenuation coefficient proportional to frequency is often a good approximation. With this variable, frequencies close to the centre frequency of interest can also be included in the plot. A linear regression on this data is then performed to give an estimate for β, which is expected to be constant in an attenuation law of the form $\exp(-\beta f R)$, where f and R are the frequency and range. The calculation is repeated for various centre frequencies, thus giving estimates of the attenuation coefficient $\alpha = \alpha(f)$, and the source signature. With this method Buchanan et al.[40] found a Q value of 45 in the Lower Florida seam, England, and Buchanan et al.[41] found a Q value of 39 in the Bulli seam, Australia.

Arnetzl et al.[49] have used a similar approach to analyse data from West Germany, although their interpretation of the results is based on a linearised version of the complex dispersion relation for SH channel waves. They obtained Q values of 95 for 'Esz-coal' and 40–60 for a more bituminous variety. Hence, measurements of Q in various parts of the world all show Q values of less than 100. A typical Q of about 60 corresponds to an attenuation of 0·45 dB per wavelength. Unfortunately, coal appears to be a rather good attenuator.

6.2. Anisotropy

One of the principal effects of anisotropy is that phase and group slowness vectors are no longer parallel. It is the *group* slowness that determines arrival times, and the *phase* slowness that appears in the generalised form of Snell's law which gives the directions of transmitted and reflected wavevectors at a boundary. Thus, if anisotropy is present and is ignored, reflection points will be incorrectly positioned.

All propagation characteristics are determined by the frequency–wavenumber ($w-k$) surface. For an isotropic, non-dispersive, two dimensional medium this surface is a circular cone. If dispersion is present the $w-k_x$ (or k_y) section is curved, and if anisotropy is present the k_x-k_y section is no longer circular. The phase slowness surface satisfies $S = k/w$, whilst the group slowness surface is given by

$$S_g = \frac{\partial k}{\partial w} \qquad (5)$$

In an anisotropic, non-dispersive medium a knowledge of either surface is sufficient to calculate the other. However, if the medium is dispersive then the group slowness may be calculated from the phase slowness, but the reverse cannot be accomplished without additional information. In ISS it is usually the group slowness that is measured. Fortunately Fermat's principle of stationary time is satisfied[40] and this can be used as the basis of calculating the true reflection points.

Buchanan *et al.*[40] have used this result to modify the DTG stacking procedure, and have conducted tests on synthetic data to explore the practical aspects of the presence of velocity anisotropy. They find that the isotropic DTG stack produces three possible errors when anisotropy is present but is ignored. Firstly, the estimated fault distance may be incorrect. Secondly, the fault may be shifted laterally along its axis. Thirdly, the fault may be slightly elongated with a small amount of curvature at the ends. It is significant that the fault is *not* rotated in the horizontal plane. Anisotropic DTG stacking removes these errors. Thus in practice it is sufficient to process all the data as if isotropic. Only a few final sections will require the more complex anisotropic stack prior to final interpretation.

6.3. Channel Wave Geotomography

In recent years, the diagnosis of medical patients' ailments has been eased by the improvement in medical imaging techniques brought about by the introduction of scanning X-ray and ultrasonic tomography. These techniques are usually implemented in transmission. Waves are propagated through an inhomogeneous space (e.g. the brain). The problem is to record a set of measurements which can be used to analyse the transmission medium. Arrival times or amplitudes are commonly used; the former can be related to velocity inhomogeneities, and the latter to attenuation. For example, the arrival time t is given by

$$t = \int_L \frac{ds}{V(x,y)} \qquad (6)$$

where L is the path followed, and $V(x,y)$ the (two-dimensional) velocity field that must be found. These integrals are called projections or Radon transforms (after Radon who first solved the problem of image construction by inversion of the transform). Herman[50] has provided a useful introduction to the subject.

There are numerous algorithms for calculating $V(x,y)$. For example, there exist generalised inverse methods for which the theory is mathematically rigorous, and this ultimately involves the computation of the generalised inverse of a very large non-square matrix. Difficulties arise with data storage, convergence, stability, round-off errors and computer time. Alternatively, extensive development has taken place in the field of algebraic reconstruction (ART),[51] in which a trial solution is iteratively changed until its projections are consistent with the measured data. ART is much less mathematically rigorous. No attempt is made to solve the problem in a single, massive calculation; instead each projection is treated separately and repeatedly. ART can be implemented in several ways.[52]

Reflection tomography can be set up mathematically but is much more difficult to implement in practice because of the complications introduced by ignorance of reflection amplitude and by multiple arrivals. Geophysical, and in particular seismic, exploration is dominated necessarily by reflection techniques. Nevertheless, there are instances where geotomography can be usefully applied. Frequently, boreholes are driven from the surface to the horizon of interest. If sensors can be located in these holes then information about the space between the holes can be gleaned. For example, Lager and Lytle[4] have used tomography to analyse electromagnetic data, shot with the purpose of mapping a burn front in a coal seam. In various parts of the world research into underground coal gasification (UCG) is being carried out. One of the factors that affects UCG is seam continuity. There is little doubt that hole-to-hole tomography can be used to establish seam structure using either seismic or electromagnetic waves.

In deep mining there is a trend towards exploration by means of long (2–3 km) roadways to block out a large part of the seam. Such exploration will be aided greatly by reflection ISS but also the geometry will provide a suitable site for channel wave tomography. Mason[53] has reported results from such an experiment. By using the arrival times of the head waves he was able to use ART to construct a map of the velocity field in the coal. The particular seam involved (the High Hazles) had been partially undermined and the resulting partial subsidence produced a non-uniform stress pattern in the High Hazles. Stress fields are known to affect seismic velocities: for example, the high stress (abutment) zone surrounding any mine roadway produces a change in seismic velocity.[19,54] In this case the stress and velocity field contours matched closely thus proving that non-uniform stress can be mapped by this technique. Faults frequently have an associated abutment zone; thus it should be possible to map fault distributions using channel wave tomography.

6.4. Alternative Sources

Although explosive seismic sources are used extensively in standard surface exploration, weight dropping and vibratory sources also have wide application. These latter sources are much more controllable and this gives advantages in the seismic sense, as well as practical advantages since they are non-destructive. It is natural therefore to ask whether similar sources have been used in ISS.

There are two problems to be faced. Firstly, the source must act on the vertical coal face rather than on horizontal ground, so that redesign is necessary. Secondly, and perhaps more seriously, such a source must be fully approved by mine safety authorities if it is used on the face itself. Such approval is not easily given, and a speedy conclusion is not the norm. It may be that the mine safety regulations degrade the performance of the source or make it impracticable by, for example, making it too heavy. It is a daunting prospect to any researcher.

In spite of these difficulties Owen and his colleagues (Suhler et al.[55] and private communication from Owen) have designed a high frequency piezoelectric transducer for generating shear waves on a coal face. They have successfully generated and recorded high order SH channel wave modes in the kilohertz range. Such a system is likely to be useful for short-range, high-resolution investigations (for example, the advance detection of water filled voids in mines).

As we have seen, one of the problems to be overcome in ISS is that of dispersion. In theory it is possible to design a source that eliminates a large proportion of the dispersion. The recorded seismic signal $s(x, t)$ can be written as

$$s(x, t) = \int dw F(w) \exp[ik(w)x - iwt] \tag{7}$$

$F(w)$ is the source spectrum, $k(w)$ is the dispersion relation, and x is the total propagation path length. For simplicity we assume a frequency independent reflection coefficient, no spreading loss and no attenuation. Let w_0 be the dominant frequency. Then

$$k(w) \simeq k_0 + \left.\frac{\partial k}{\partial w}\right|_{w_0} (w - w_0) + \frac{1}{2}\left.\frac{\partial^2 k}{\partial w^2}\right|_{w_0} (w - w_0)^2 \qquad (8)$$

In this Taylor expansion it is the third (and higher terms) that are responsible for the signal spreading out as it propagates. The first two terms are related to the phase and group velocities. We re-write the third term as $k_2(w - w_0)^2$. In radar parlance k_2 is the chirp rate. So after propagating the distance x the signal is chirped by an amount $k_2(w - w_0)^2 x$ (neglecting higher terms). Suppose now the source spectrum is

$$F(w) = A(w) \exp[-iK(w - w_0)^2] \qquad (9)$$

then the total chirp that the signal $A(w)$ suffers is $(k_2 x - K)(w - w_0)^2$. Thus if $K = k_2 x$ the chirp is zero. If the seam propagation characteristics have been investigated previously then k_2 is known. Thus if a source can be constructed to emit a signal of the (real) form of eqn 9 together with a *switchable* chirp K then it can be turned selectively to any range x. Physically this means that the source emits a dispersed signal which gradually compresses as it propagates until it has travelled a distance x, after which it would disperse. This method of course also suffers from the operational difficulties raised earlier.

An alternative is to ask what equipment, already in use in the mine, might be used as a seismic source. There are several candidates such as drilling machines, but the obvious one is the coal cutter itself. The energy of a cutter is dissipated in several ways, but for the present purposes, it may be viewed as a generator, perhaps inefficient, of seismic energy. The cutter tracks across the face; hence it is possible to locate a few geophones on the face and record seismograms for various cutter locations. This situation is similar to a synthetic aperture radar. The potential use of the cutter as a source ultimately depends on whether or not it emits an identifiable seismic signal, and Buchanan et al.[56] have shown that it does, although considerable processing is required. The method has not been used routinely but certainly has potential advantages. Unlike other techniques, production is not interrupted for the period of the survey; indeed, production is an integral part of the survey method. However, interruption is only a problem on those few faces where coal is mined continually.

7. CONCLUSIONS

The principal conclusion is that reflection in-seam seismics can be used as a routine aid to mine planning. The technique has been fully demonstrated in several countries to the satisfaction of the users, the mining engineers. In the UK and elsewhere the method has been added to the arsenal of techniques at the disposal of planners, and it is as readily available as any other.

Because of the mining methods predominant in Europe, research has been concentrated preferentially on reflection ISS, but in other parts of the world transmission surveys may be more useful. In particular hole-to-hole geotomographic surveys can be used to aid planning of strip mines, underground coal gasification plants and shallow underground mines. Routine exploitation has not quite been reached but very soon will be.

Like any other new technique there is ample scope for further developments. On the acquisition side, the time is surely ripe for the development of a controlled source. A less bulky, low-powered recording system would also be advantageous. Currently we are able to give management the location of discontinuities in the seam. Sometimes we can estimate the size of faults (face stopping or not) on the basis of past experience. There is no doubt, however, that management would like information on the nature (fault, washout, dyke, etc.) of discontinuities, and on the size (throw, hade, etc.). Such information must be at least partially encoded in the reflected signals. The outstanding processing or interpretation problem to be solved is that of 'target recognition'. It is likely that the solution will come from a more complete theoretical understanding of the scattering of channel waves from such structures, and also from a statistical pattern recognition analysis of previous results which have been verified by mining. Perhaps the most difficult problem if all is to obtain funding from the controllers of the purse strings, for in these difficult economic times they are wont to accept what has been developed and to dismiss the need for improvement. Such a view is shortsighted.

ACKNOWLEDGEMENTS

Over the years a large number of persons have been involved in the National Coal Board's programme to develop in-seam seismics, and it is impossible to name them all. However, I am particularly grateful to colleagues at MRDE, especially Vaughan Thomas, Peter Jackson, Paul

Taylor and Richard Davis. Good advice on the problems of mining was provided by Mike Clarke and Allan Callis. I have enjoyed a long and fruitful collaboration with Iain Mason of Oxford University, and some of his students, particularly Tony Booer, provided valuable insight. From time-to-time John Hudson of Cambridge University kept me on the right path with regard to theory. Field work would be impossible without the willing cooperation of colliery personnel, and the advice and assistance of many colliery managers, geologists, surveyors and their staff is greatly appreciated; Bob Hoare, especially, provided much impetus here. Finally, sincere thanks must go Peter Tregelles, Director of MRDE, who had the faith to persevere and obtain funding even when results were thin on the ground.

The views expressed in this chapter are those of the author alone, and do not necessarily represent those of the National Coal Board.

REFERENCES

1. EZRA, SIR DERFK (1978) *Coal and Energy*, Ernest Benn, London, p. 14.
2. ZIOLKOWSKI, A. M. (1979) Seismic profiling for coal on land. In *Developments in Geophysical Exploration Methods—1*, Ed. A. A. Fitch, Applied Science Publishers, London, p. 271.
3. HASBROUCK, W. P. and HADSELL, F. A. (1976) Geophysical exploration techniques applied to Western United States coal deposits. In *Coal Exploration Proceedings of the First International Coal Exploration Symposium*, London, Ed. W. L. G. Muir, Miller Freeman, San Francisco, p. 256.
4. LAGER, D. L. and LYTLE, R. J. (1977) Application of ART algorithms for defining a subsurface electrical profile from high frequency measurement. *Radio Science* **12**, 2.
5. DAVIS, D. T., LYTLE, R. J. and LAINE, E. F. (1978) High-frequency electromagnetic wave probing of an *in situ* process. Paper presented at the 48th SEG Meeting, San Francisco.
6. LAINE, E. F. (1980) Detection of water-filled and air-filled underground cavities. Lawrence Livermore Laboratory, paper UCRL-53127.
7. REES, D. H. Private communication.
8. SUHLER, S. A., OWEN, T. E., HIPP, J. E. and PETERS, W. R. (1978) Development of a deep penetrating borehole geophysical technique for predicting hazards ahead of coal mining. Final technical report prepared by Southwest Research Institute for the US Department of Interior, Bureau of Mines, Denver.
9. ZIOLKOWSKI, A. M. (1976) High resolution seismic reflection developments in UK coal exploration. *Proceedings of the Coal Seam Discontinuities Symposium*, Pittsburgh.
10. ZIOLKOWSKI, A. M. and LERWILL, W. E. (1979) A simple approach to high resolution seismic profiling for coal. *Geophysical Prospecting* **27**, 360–93.

11. COON, J. B., REED, J. T., CHAPMAN, W. L. and DUNSTER, D. E. (1978) Surface seismic methods applied to coal mining problems. Paper presented at the 48th SEG Meeting, San Francisco.
12. RÜTER, H. (1978) An introduction to the problems of the Ruhr area. Paper presented at the 48th SEG Meeting, San Francisco.
13. COOK, J. C. (1975) Radar transparencies of mine and tunnel rocks. *Geophysics* **40**, 865–85.
14. FOWLER, J. C. and HALE, S. D. (1980) Coal seam hazard detection using synthetic-pulse radar. *Proceedings of the 50th SEG Meeting*, Vol. 5, Houston, p. 3121.
15. EVISON, F. F. (1955) A coal seam as a guide for seismic energy. *Nature* **176**, 1224–5.
16. KREY, T. C. (1963) Channel waves as a tool of applied geophysics in coal mining. *Geophysics* **28**, 701–14.
17. MASON, I. M., BUCHANAN, D. J. and BOOER, A. K. (1980) Fault location by underground seismic survey. *IEE Proc. Part F* **127**, 322–36.
18. BUCHANAN, D. J. (1978) The propagation of attenuated SH channel waves. *Geophysical Prospecting* **26**, 16–28.
19. MASON, I. M., BUCHANAN, D. J. and BOOER, A. K. (1980) Channel wave mapping of coal seams in the United Kingdom. *Geophysics* **45**, 1131–43.
20. LAGASSE, P. E. and MASON, I. M. (1975) Guided modes in coal seams and their application to underground seismic surveying. *Proc IEEE Ultrasonic Symposium*, Los Angeles, pp. 64–7.
21. FRANSSENS, G., LAGASSE, P. E. and MASON, I. M. (1980) The leaking shear horizontal modes of in-seam exploration seismology. *Proceedings of the 50th SEG Meeting*, Vol. 5, Houston, pp. 3045–67.
22. LAGASSE, P. E. (1975) Finite element analysis of piezoelectric elastic waveguides. *IEEE Trans Sonics and Ultrasonics*, SU-20, 354.
23. GUU, J. Y. (1975) Studies of seismic guided waves: the continuity of coal seams. Ph.D. Thesis, Colorado School of Mines, Golden, Colorado.
24. YANG, C. H. (1976) Elastic waves: P & S conversions in first arriving reflections and critical refractions. Ph.D. Thesis, Colorado School of Mines, Golden, Colorado.
25. SU, F. C. (1976) Seismic effects of faulting in coal seams: numerical modelling. Ph.D. Thesis, Colorado School of Mines, Golden, Colorado.
26. DRESEN, L. and FREYSTÄTTER, S. (1976) Rayleigh channel waves for the in-seam seismic detection of discontinuities. *J. Geophys.* **42**, 111–29.
27. DRESEN, L. and FREYSTÄTTER, S. (1978) The influence of oblique-dipping discontinuities on the use of Rayleigh channel waves for the in-seam seismic reflection method. *Geophysical Prospecting* **26**, 1–15.
28. FREYSTÄTTER, S. and DRESEN, L. (1977) Propagation of Rayleigh channel waves in coal seams—model seismic investigations. *J. Geophys.* **43**, 807–28.
29. DRESEN, L., KERNER, C. and KÜHBACH, B. (1981). The influence of an asymmetrical sequence 'rock-coal-rock' on the propagation of Rayleigh seam waves. Presented at the 43rd EAEG Meeting, Venice.
30. BOOER, A. K. (1982) Underground geophysics of coal seams. In *Developments in Geophysical Exploration Methods—3*, Ed. A. A. Fitch, Applied Science Publishers, London, pp. 1–32.

31. KLAR, J. W. P. and ARNETZL, H. H. (1978) A new firedamp-proof instrument for in-seam seismics in coal mining. Paper presented at the 40th EAEG Meeting, Dublin.
32. RÜTER, H. and SCHEPERS, R. (1981) An intrinsically safe data acquisition system for seismic measurements in coal mines. Paper presented at the 43rd EAEG Meeting, Venice.
33. BUCHANAN, D. J. (1979) The location of faults by underground seismology. *Colliery Guardian* **227**, 419–28.
34. BUCHANAN, D. J., DAVIS, R., JACKSON, P. J. and TAYLOR, P. M. (1981) Fault detection in coal by channel wave seismology: some case histories. *Bull. Australian SEG* **12**, 13–19.
35. THOMSON, W. T. (1953). Transmission of elastic waves through a stratified solid medium. *J. Appl. Phys.* **21**, 89.
36. HASKELL, N. A. (1953). The dispersion of surface waves on multilayered media. *Bull. Seis. Soc. Am.* **43**, 17–34.
37. MCMECHAN, G. and YEDLIN, M. J. (1981) Analysis of dispersive waves by wave field transformation. *Geophysics* **46**, 869–74.
38. MILLAHN, K. O. and ARNETZL, H. H. (1979) Analysis of digital in-seam reflection and transmission surveys using two components. Paper presented at the 41st EAEG Meeting, Hamburg.
39. DZIEWONSKI, A. M., BLOCH, S. and LANDISMAN, M. (1969) A technique for the analysis of transient seismic signals. *Bull. Seis. Soc. Am.* **59**, 427–44.
40. BUCHANAN, D. J., JACKSON, P. J. and DAVIS, R. (1983) Attenuation and anisotropy of channel waves in coal seams. *Geophysics* **48**, 133–47.
41. BUCHANAN, D. J., JACKSON, P. J., TAYLOR, P. M. and DOYLE, J. F. The application of the in-seam seismic method in Australian coal mines. *Proc. Aust. Inst. Mining and Metallurgy* (in press).
42. BUCHANAN, D. J. and JACKSON, P. J. (1983) Dispersion relation extraction by multitrace analysis. *Bull. Seis. Soc. Am.* **73**, 391–404.
43. BERESFORD-SMITH, G. and MASON, I. M. (1980) A parametric approach to the compression of seismic signals by frequency domain transformation. *Geophysical Prospecting* **28**, 551–71.
44. MARSCHALL, R. and SCHOTT, W. (1981) Treatment of dispersive wave trains. Paper presented at the 43rd EAEG Meeting, Venice.
45. BUCHANAN, D. J., DAVIS, R., JACKSON, P. J. and TAYLOR, P. M. (1981) Fault location by channel wave seismology in United Kingdom coal seams. *Geophysics* **46**, 994–1002.
46. MILLAHN, K. O. and MARSCHALL, R. (1980) Two-component in-seam seismics. *Proceedings of the 50th SEG*, Vol. 5, Houston, pp. 3019–48.
47. MILLAHN, K. O. and KNECHT, M. (1981) Two-component in-seam seismics. Paper presented at the 43rd EAEG Meeting, Venice.
48. BUCHANAN, D. J., DAVIS, R., JACKSON, P. J. and TAYLOR, P. M. (1981) The use of channel wave seismology to find faults in coal seams. *Proceedings of the 50th SEG Meeting*, Vol. 5, Houston, pp. 2985–3018.
49. ARNETZL, H. H., KNECHT, M. and KREY, T. (1982) Theoretical and practical aspects of absorption in the application of in-seam seismic coal exploration. *Geophysics* **47**, 1646–56.
50. HERMAN, G. T. (1979) *Topics in Applied Physics, Image Reconstruction from*

Projections: Implementation and Applications, Vol. 32, Springer-Verlag, Berlin.
51. GORDON, R. (1974) A Tutorial on ART, *IEEE Trans. Nuc. Sci.* **NS-21**, 78–93.
52. SHEPP, L. A. and KRUSKAL, J. B. (1978) Computerized tomography: the new medical X-ray technology. *American Mathematical Monthly* **85**, 420–39.
53. MASON, I. M. (1981) Algebraic reconstruction of a two dimensional seismic channel wave velocity field in the High Hazles seam at Thoresby Colliery. *Geophysics* **46**, 298–308.
54. PROSKURYAKOV, V. M., KRYZHANOVSKII, M. V. and SMIRNOV, V. A. (1976) Seismic investigation of abutment pressure ahead of the coalface. *Ugol. Ukrainy*, No. 9, 18–19.
55. SUHLER, S. A., DUFF, B. M., OWEN, T. E. and SPIEGEL, R. J. (1978) Geophysical hazard detection from the working face. Southwest Research Institute contract H0272027 for the US Bureau of Mines.
56. BUCHANAN, D. J. MASON, I. M. and DAVIS, R. (1980) The coal cutter as a seismic source in channel wave exploration. *IEEE Trans. Geoscience and Remote Sensing* **GE-18**, 318–20.
57. BAKER, G. A. (1975) *Essentials of Padé Approximants*, Academic Press, New York.

Chapter 2

UNDERGROUND GRAVITY SURVEYS

A. HUSSAIN

Digital Exploration Ltd, East Grinstead, Sussex, UK

SUMMARY

The gravimeter used in underground mines is a three-dimensional exploration device which provides very valuable information about the density distribution around the measurement level. In this article we will examine its application for density determination in situ and for 'anomalous zone' definition in the mines. Practical examples of subsurface gravity surveys carried out in the shafts of a coal mine and in different levels of a lead–zinc mine will be presented. Gravity anomaly maps and profiles were constructed from different levels and compared with the known geology. The results demonstrate that gravity studies are the best first phase of study for defining drilling targets.

1. INTRODUCTION

Basic to the gravity prospecting method is the density contrast among different rock types which generate gravity anomalies. Surface gravity surveys provide information about the lateral density variations below the level of measurements. Subsurface gravity measurements, on the other hand, furnish a three-dimensional mass distribution picture which makes the gravity method an important observation for mineral and hydrocarbon exploration.[1-5] Underground gravity measurements can be conducted vertically in the boreholes and in the mine shafts and horizontally in different levels of the mines.

The vertical gravity measurements have been proven to be very important geophysical tools with several applications in prospecting. The major advantage of using a gravimeter in this way is that a large volume of rock through which the measurements are made affects the gravity readings. About 90% of the gravity effect is derived from within five times the vertical separation between the measurements assuming horizontal and homogeneous lithologic units.[6] Therefore, the gravimeter is used as a three-dimensional sensing device and is potentially the most promising method for determining densities *in situ* of a large volume of rock, whereas the other density logging systems (gamma-ray) have a limited radius (the order of tens of cm) of investigation and provide only a one-dimensional density distribution along the walls of the borehole.

Gravimetrically determined densities have a number of applications. Porosity can be determined from the density if the fluid content and saturation are known which is useful in production and reservoir studies.[7] Moreover, density data can be used to determine zones of gas saturation. Its most important use, however, is its aid in the interpretation of other geophysical data. Precise vertical gravity measurements can lead to accuracies of within 0.01 g/cm^3 in density and 0.05% in porosity determinations.

In most of the mining districts the ore bodies and related structures are irregularly distributed and lie mostly under thick overburden occurring at depths beyond the resolution of surface geophysical surveys. With the increased demand for minerals as raw material and with improved mining techniques, areas of considerable depths (1000–1500 m) have become targets for further exploitation. This tendency to explore for new ore at these depths requires special techniques. The existing galleries in the underground mines offer the chance to move the geophysical prospecting methods nearer to the target.

Theoretically, several geophysical methods can be applied in underground mines but some of these methods such as electrical, electromagnetic and self potential methods very often do not yield satisfactory results. Any of these methods will locate ore bodies only if certain geological and physical conditions are fulfilled. For example, iron pipes, railroads and other iron material and electric cables in the galleries render electromagnetic methods impossible. Resistivity, induced polarisation and self potential methods require specific conditions to provide good results.

The gravity method, on the other hand, is based only on the density contrast between different rock types and mineralised areas and can be successfully applied in these conditions. Where there are adequate density

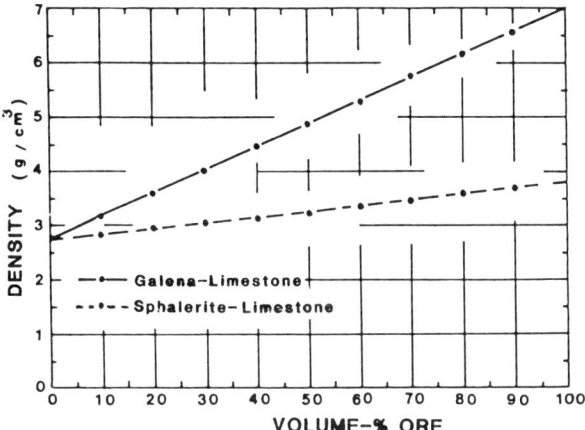

FIG. 1. Plot showing the relationship of density and percentage (volume) of sulphide ore within limestone matrix of density 2.72 g/cm^3.

contrasts between mineralised zones and the surrounding areas, gravity studies are the best and least expensive first phase of study for 'anomalous zone' definition. In addition, because of the structural and lithological controls on mineralisation, major geological structures and lithologic boundaries may be inferred from the gravity data.

Most ore bodies and associated rock alterations possess densities differing considerably from the host rock. The density of the host rock is mostly uniform, whereas within the mineralised zones it may be increased by the presence of minerals. Figure 1 illustrates the theoretical relationship between density and volume percent of galena and sphalerite in a limestone matrix; both of these are commonly approximated in nature. In this example, a density of 7 g/cm^3 was used for galena, 3.80 g/cm^3 for sphalerite and 2.72 g/cm^3 for limestone. For example, for a volume of rock containing 4% galena in limestone matrix, a density contrast of 0.17 g/cm^3 can be expected. This will be the maximum density contrast. The high porosity of ore or a matrix of different lithology will, however, change this relationship.

A region or zone in which the density is unusually high or low constitutes a gravity anomaly. In underground surveys both negative as well as positive anomalies are of interest as they can be caused by the ore bodies. Whether the type of anomaly is positive or negative depends on the density of the anomalous zone and its location with respect to the level of measurements. A positive anomaly may mean a dense material below the level of measurement or a deficiency of mass above it or a combination of both situations.

One of the difficulties in underground survey is that measurements can be carried out only in the existing galleries. Owing to this difficulty the results of this kind of survey do not have to be treated in the same way as the surface surveys. The gravity data obtained from different levels of the mine are often not sufficient to permit an accurate quantitative interpretation. But qualitatively, with the careful consideration of surrounding geology, it is possible to mark the anomalous density zones. The purpose of this paper is to discuss and present some case histories illustrating the correlation of mineralised zones and related features to gravity anomalies, showing at the same time the applications and limitations of this method to underground prospecting.

2. VERTICAL GRAVITY MEASUREMENTS

Vertical gravity measurements provide us with the possibility of very accurate density determination *in situ* of a large volume of rock through which measurements are carried out. These densities are obtained from the change in gravity over a known vertical interval within the earth. The gravity difference $\Delta g_{i,i+1}$ over a vertical interval $\Delta H_{i,i+1}$ in the subsurface consisting of a homogeneous horizontal layer of density δ is given by:

$$\Delta g_{i,i+1} = (F - 4\pi G \delta)\Delta H_{i,i+1} + \Delta T_{i,i+1} \qquad (1)$$

where F is the free air vertical gradient of gravity, G is the universal gravitational constant and $\Delta T_{i,i+1}$ is the variation in the topographic correction over the vertical interval $\Delta H_{i,i+1}$.

By solving eqn (1) for the density

$$\delta = F/4\pi G - (\Delta g_{i,i+1} - \Delta T_{i,i+1})/4\pi G \Delta H_{i,i+1} \qquad (2)$$

and substituting a normal value of $F = 0{\cdot}3086\,\text{mgal/m} = 3{\cdot}086 \times 10^{-3}\,\text{s}^{-2}$ and $G = 6{\cdot}67 \times 10^{-8}\,\text{cm}^3/(\text{g s}^2)$, we have

$$\delta(\text{g/cm}^3) = 3{\cdot}687 - 11{\cdot}93(\Delta g - \Delta T)_{i,i+1}/\Delta H_{i,i+1} \qquad (3)$$

In this formula the assumption is made that the infinite lithologic units are horizontal, homogeneous and uniformly thick. Possible effects from anomalous masses located below or above the vertical interval and the departure from the infinite slab model are not considered here, but distortion in the densities due to the above-mentioned factors have been documented in the literature.[8-10]

The vertical gravity readings are corrected for the instrumental drift and the tidal effects. If the measurements are made in the mine shafts (in the cage) or in the galleries instead of boreholes, an extra correction for this mass deficiency is also necessary. The shaft correction can be calculated by computing the effect of an equivalent vertical cylinder and the gallery (mine drift) correction can be computed by assuming its two-dimensionality and a suitable geometrical shape.[5] The gallery effect can be significantly reduced by mounting the gravimeter on a high tripod and placing it away from the walls so that the mass deficiency below and above the level of measurements cancels out. The same is true for the shaft effect where the correction is almost negligible a few metres away from the shaft ends.

The most important correction in the areas of high topographic relief is the terrain correction. It can attain a considerable value depending not only on the elevations of the surface topographic features but also on the relative position of the underground stations. Thus, this correction may increase or decrease with depth depending on whether there is an excess or a deficiency of mass with reference to the terrain datum. For the topography lying above the terrain datum a positive correction is applied and for topography below it a negative correction is necessary.

Figure 2 shows a plot of topographic correction versus depth at two different locations of a mining district situated in extremely rugged topography. This correction was computed by the method described by Hearst[11] and Hussain et al.[5] The topography within a 20 000 m radius of the shafts was levelled on a terrain datum passing through the surface points of the shafts. The magnitude of the correction for the surface points of both shafts is almost identical. In the case of the West shaft it increases rapidly with depth, whereas in the second case (the Stefani shaft) the topographic effect decreases with depth and attains a value of 5·5 mgal at a depth of 650 m as compared to 15 mgal in the first case at the same depth. However, the common feature in both cases, is that the curves become smoother with depth.

This damping effect is due to the greater vertical distance of the topographic features, which provides a definite advantage of subsurface gravity survey over surface work and provides a possibility for exploration in areas of rough terrain where work on the surface for this purpose has practically no value.

The error in densities determined gravimetrically, assuming a negligible error in topographic correction and in vertical distance measurement, is mainly controlled by the precision of the gravity measurements and the length of the vertical interval. By differentiating eqn (2) with respect to Δg

40 A. HUSSAIN

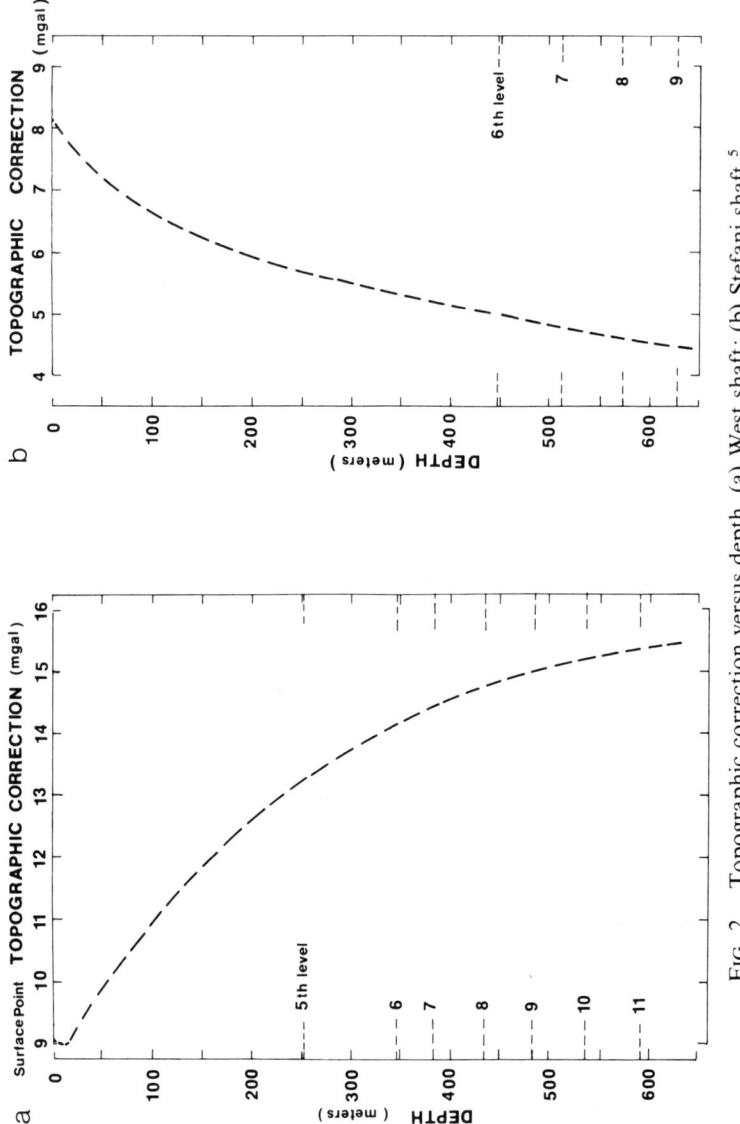

FIG. 2. Topographic correction versus depth. (a) West shaft; (b) Stefani shaft.[5]

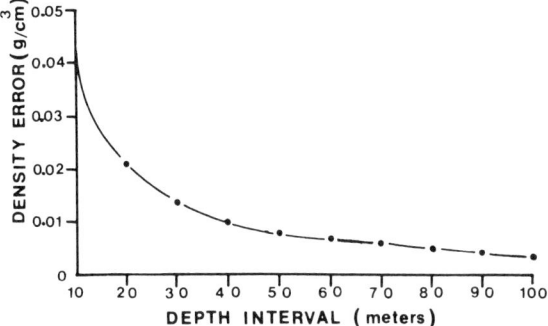

FIG. 3. Plot of the density error versus depth intervals for an accuracy of 0·03 mgal.[5]

and solving for density error ($\Delta\delta$) in terms of gravity error ($\Delta\Delta g$) and interval distance (ΔH), we obtain[12]

$$\delta\Delta = -\Delta\Delta g/(0\cdot0836\,\Delta H)$$

which shows that the density error is directly proportional to the gravity error and inversely proportional to the vertical separation of the measurements (ΔH). Figure 3 shows a plot of the density error drawn as a function of the vertical (depth) interval for a precision of 0·03 mgal in gravity readings. With precise gravity measurements it is possible to achieve an accuracy better than 0·01 g/cm^3 in density determinations.

2.1. Applications

In this section, examples of vertical gravity measurements in five shafts of a coal and a lead–zinc mine situated in the central Alps will be presented. The first example (Fig. 4) deals with the gravity logs (density) from the three shafts of a deep coal mine situated in a Tertiary basin. The basement of the basin is crystalline mica schists and amphibolite schists. A geological section through two of the shafts is also shown in the upper part of the figure. The lower most sedimentary rocks are sandstone overlain by sandy, partially carbonatic marl, which is overlain by a 20–60 m-thick layer of Quaternary gravels. The coal seam in this region has an average thickness of 6 m and occurs in the marl and sandstone boundary.

The gravity measurements were made using an exploration LaCoste and Romberg Model G gravimeter. Measurements directly in the shafts were not possible because of non-stop mining operations; therefore, measurements were made in the drifts connecting the shafts at different horizons by

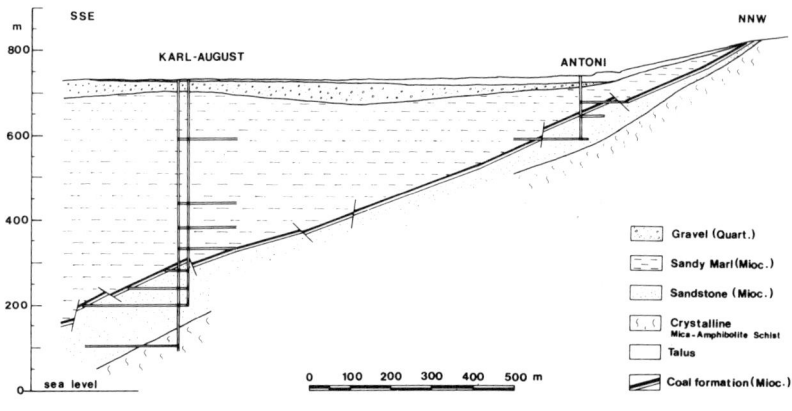
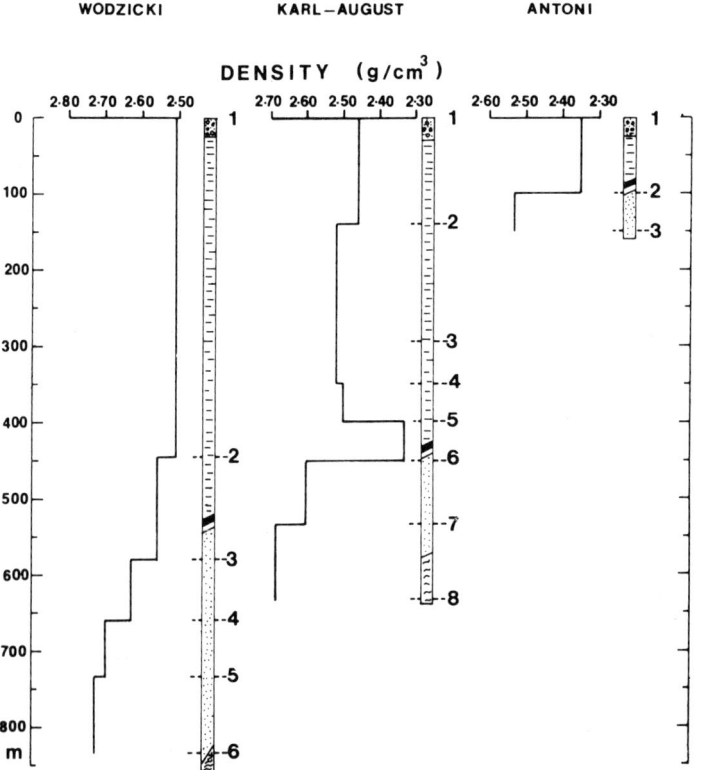

FIG. 4. Vertical density profiles for the Wodzicki, Karl–August and Antoni shafts from gravity measurements. A geological section through the Karl–August and Antoni shafts is also illustrated.[13]

placing gravimeters at an 8 m horizontal distance from the shaft centre at all the levels. Gravity readings were taken in two runs in each shaft and corrected for the effects discussed earlier.

The Antoni, Karl–August and Wodzicki shafts are 150, 635 and 833 m deep, respectively, and the densities were determined at varying intervals in each shaft. The first interval of the Antoni and Karl–August shafts represents a combined density of gravel and marls. After compensating for the effect of marl, the gravel density was determined as $2 \cdot 10 \, g/cm^3$. The marl has a fairly uniform density of $2 \cdot 52 \, g/cm^3$ and there is almost no change in its density either laterally or vertically. The most interesting feature of the density log in the Karl–August shaft is that the coal formation (6 m coal seam and about 44 m light marl containing small coal bands; density, $2 \cdot 32 \, g/cm^3$) produces a significant density contrast with the upper lying marl and lower lying sandstone (density, $2 \cdot 63 \, g/cm^3$). The density between the 2nd and 3rd levels of the Wodzicki shaft was computed as $2 \cdot 56 \, g/cm^3$ which is a combined density of marl, coal formations and sandstone. Its lower part (3–6 levels) represents the sandstone and amphibolite–biotite schist density. The density and velocity contrast between these rock types shows that relatively good reflection energy can be expected from the boundaries of a coal formation.[13]

These density results furnished information about the 3-dimensional density distribution within the basin which was helpful in interpreting other geophysical data from this area. A precision of 0·04 mgal was achieved in this survey. The minimum vertical interval in the three measured shafts was about 50 m which resulted in a density error of about $0 \cdot 01 \, g/cm^3$. In all other cases, the length of the vertical interval was more than 50 m, and therefore the accuracy of density is even better than $0 \cdot 01 \, g/cm^3$.

The next example of a gravity log is taken from a lead–zinc mine located in a tectonically disturbed area. Figure 5 shows the results of gravimetric densities compared with the hand sample densities from the shaft. The hand samples were collected from different levels within a radius of 200 m from the shaft. The hand samples show a large scatter in density. Usually it is very difficult and time consuming to get statistically representative density data for an area, because the hand samples from relatively homogeneous layers often give different density values depending upon the origin of the samples within the lithologic layers. However, the average hand sample density compares very well with the gravimetric densities except in the last two levels where the galleries were not uniformly distributed and only one part of the mine is represented.

The average schist, limestone and dolomite hand sample densities from

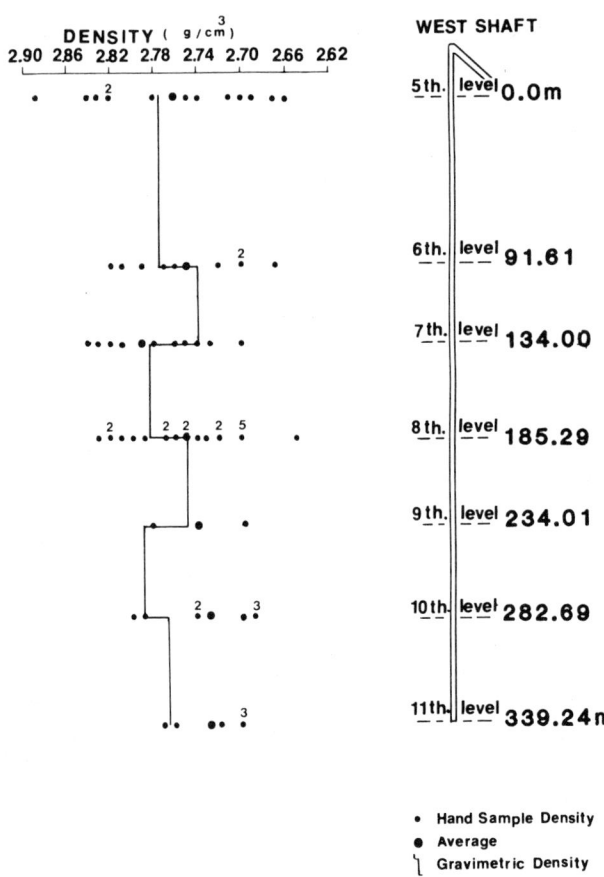

FIG. 5. Vertical density profile for the West shaft from gravity measurements compared with densities obtained from laboratory measurements on rock samples.[5]

the shaft surroundings were 2·66, 2·72 and 2·80 g/cm^3, respectively. None of these densities is represented individually in the density log, but the gravimetric densities can be seen as values which are weighted averages of the densities of these rock types. The area has been subjected to strong tectonic movements and in such cases it is possible only to obtain a combined effect of different rock formations. The density log is mainly affected by a major fault, the 'Union Kluft' which cuts down through all levels of the mine at different distances (from tens of metres to 250 m) from the shaft. This fault is the main boundary between limestone (density, 2·72 g/cm^3) and dolomite (density, 2·80 g/cm^3). The fault occurs up to 120 m from the shaft at the 5th, 8th, 10th and 11th levels and at the 6th, 7th and 9th levels it lies up to 250 m eastward from the shaft. The interval densities follow the same pattern, i.e. where the fault is nearer to the shaft (means closer dolomite), the gravimeter densities are higher (2·76–2·78 g/cm^3) and where the distance of the dolomite occurrence is more than 200 m from the shaft, the interval densities are lower (2·74 g/cm^3, since the rocks close to the shaft are mainly limestone).

3. HORIZONTAL GRAVITY SURVEYS IN THE MINES

Underground gravity surveys in the mine levels can be carried out in the same way as a routine survey of the surface, the difference being that the targets are rather small.[14] Therefore, accuracy standards for such types of surveys should be set in an efficient and cost-effective way, so as to ensure a signal-to-noise ratio large enough to adequately 'see' the targets. Repeat readings should be taken as frequently as possible which would improve the data quality and add confidence to the estimate of the data accuracy.

Most of the corrections required on the raw data can be computed in the same manner as for surface gravity data. The Bouguer correction is $4\pi\delta GH$; this is twice that on the surface. A positive correction for the mined-out areas and drifts lying below the measurement level is applied and a negative correction is necessary for such mass deficiency lying above the gravity station. The effect of galleries and mined-out areas approaches a constant value very rapidly as the vertical distance to the station increases.

In areas of rough terrain the main limiting factor on accuracy and the most time consuming is the topographic correction. A speedy, convenient and accurate method for computing this correction for underground stations is the preparation of a digital terrain model of the mining district, composed of suitably sized prisms (depending upon topography). Once this model is prepared it is very simple to compute the topographic correction for as many stations as required at any level of the mine by reducing the topography to a horizontal plane passing through the reference level in question. As the topographic effect is smoother at depth, the amount of work can be reduced by calculating terrain correction on a suitable grid (depending upon the terrain and the station spacings) and interpolating for the primary gravity stations.

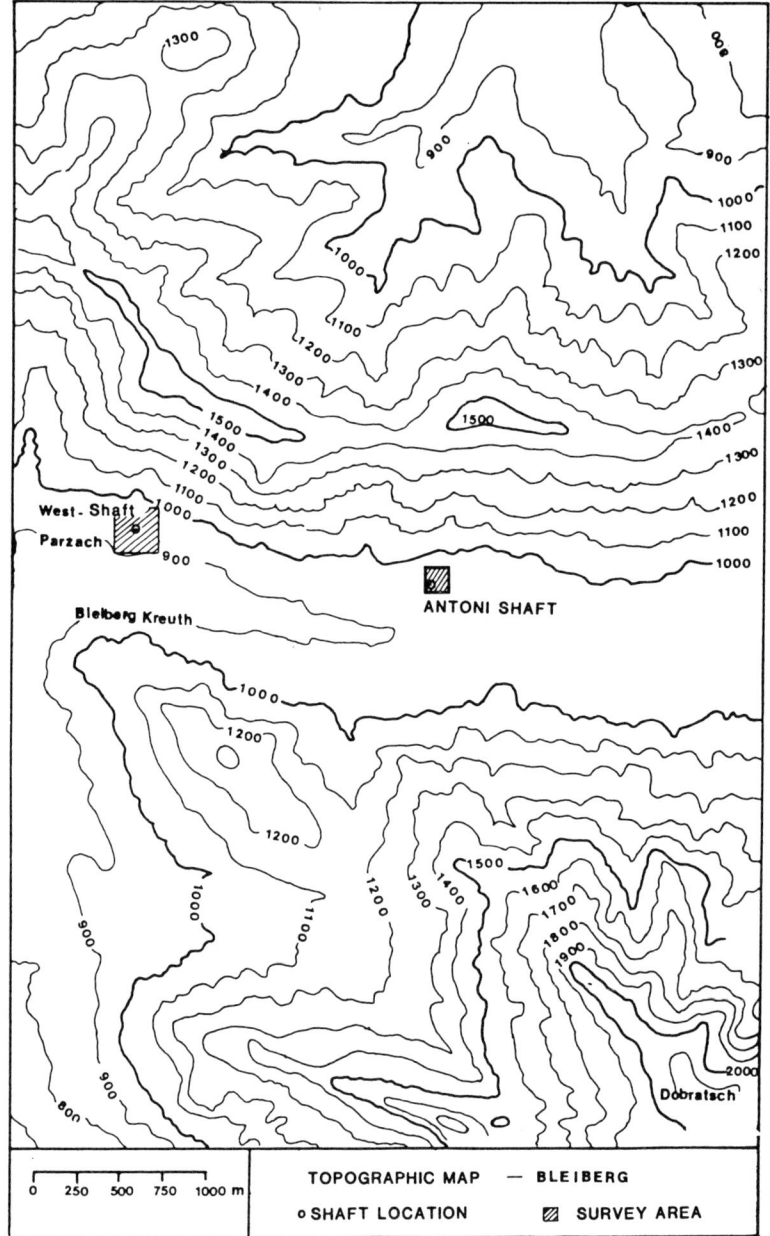

FIG. 6. Topographic map of the survey area. Contour interval, 100 m.

Underground gravity data, after making necessary corrections, can be presented in two ways:

In the form of vertical sections of gravity data.
In the form of Bouguer anomaly maps and profiles at each level.

The first method of data presentation is very useful because the gravity gradients in the anomalous areas are much stronger than those found on the surface fields. The measured gravity gradients can be interpreted in terms of apparent density derived for different geometrical shapes from a simplification of Poisson's equation describing the potential field within the gravitating matter.[8] However, in the case where the measurement of vertical gradients is not possible, the data presentation is made through the gravity anomaly maps and profiles.

As the gravity survey can be carried out only in the existing galleries, the data is insufficient, mostly, to permit quantitative interpretation. But qualitatively, with the careful consideration of the surrounding geological setting, it is possible to mark the anomalous density zones.

3.1. Examples of Gravity Survey

An underground gravity survey was carried out in different levels of the lead–zinc mine in Bleiberg (Carinthia, South Austria). The mining district lies in the Triassic calcareous Alps. The underground mine, lying in extremely rugged topography (Fig. 6), reaches a depth of 750 m below the surface. The out-cropping rocks are dolomite, limestone, schist and sandstone. The elevation ranges from 800 m to 2200 m over a distance of 5 km from the survey area. A total of about 410 gravity stations were occupied on eight different levels (six around the West shaft and two around the Stefani shaft) of the mine.

The stratigraphy of the Bleiberg unit (Fig. 7) is well known from the mine workings.[15,16] The hanging wall of the Triassic sequence ends with the Noric main dolomite ranging in thickness from 1000–1500 m. This is followed by the Karnian which displays a differential sequence of sedimentation of the Triassic and includes three to four layers of schists 3–25 m in thickness interbedded with calcareous rocks. The Ladinian represents a limestone–dolomite sequence of 750–850 m. The hanging wall parts of these beds were investigated in detail because of the significance of the sedimentation of the content in ore minerals. Lately, a reef facies became known in addition to the well-stratified beds.[15] Of the 2300–3000 m of the sediments in the Bleiberg region, only a 140 m cross-section

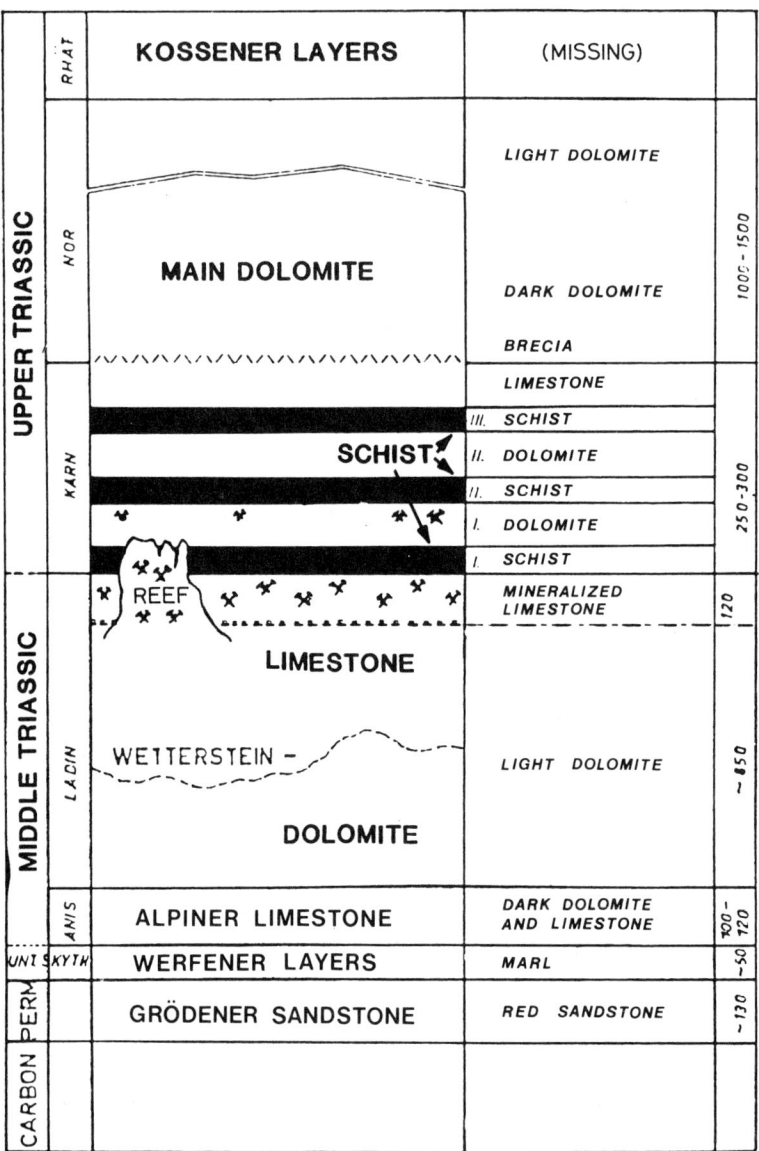

FIG. 7. Stratigraphy of Bleiberg area.[15]

(marked with mining insignia in Fig. 7: 120 m in the Wetterstein limestone and about 20 m in the dolomite between the schist layers) is actively productive of galena and sphalerite. The ores occur both in stratigraphically controlled fissures and in beds (vein type and interbedded).

The area has been subjected to severe tectonic movements and, therefore, the ore bodies are irregularly distributed. Long galleries must be driven before a new ore body is discovered. It may happen, however, that an ore body lies only a few metres away from the drift and remains undiscovered. As drilling costs are very high, the need for a method to define drilling targets is essential for efficient exploration. An experimental gravity survey was carried out to study its applications and limitations in underground prospecting in topographic and tectonic situations such as Bleiberg. Gravity results of each level will be discussed one by one in the following sections.

3.1.1. Level 5

The geological setting of the 5th level (251 m subsurface) of Bleiberg Mine around the West shaft is shown in Fig. 8 where the locations of gravity stations are also marked. The hand sample densities are marked on the sample locations. These densities were determined to compute a density model of this part of the mine as an extra aid for the interpretation of gravity data. The geology is mapped and interpreted from the intensive mining activities in this mining district. The major fault, the 'Union Kluft', as discussed in Section 2.1, is the main boundary between dolomite and limestone. The rock density is $2 \cdot 67 \text{ g/cm}^3$ near the shaft but increases to about $2 \cdot 82 \text{ g/cm}^3$ near the fault in gallery 'A'. The same situation can be observed in the upper part of the map. The average density in area 'D' is about $2 \cdot 70 \text{ g/cm}^3$ whereas only 40 m west in 'B' the density is over $2 \cdot 80 \text{ g/cm}^3$. The density near Union Kluft is low because of the presence of a mylonitic zone.

Keeping in mind the complicated geological situation and the density contrast zones, it became evident that it would be necessary to make observations at a maximum distance of 10 m if one wished to define the gravity picture in detail. Shaft stations on each level were established as the base stations (using a LaCoste and Romberg Model G gravimeter) for the level in question and finally tied to the main base which was established previously in a quiet area of the mine.

To obtain accurate data, the base station was occupied after every four stations and 15% of gravity stations were repeated. This procedure was adopted for all the levels which increased confidence in the data. Elevations

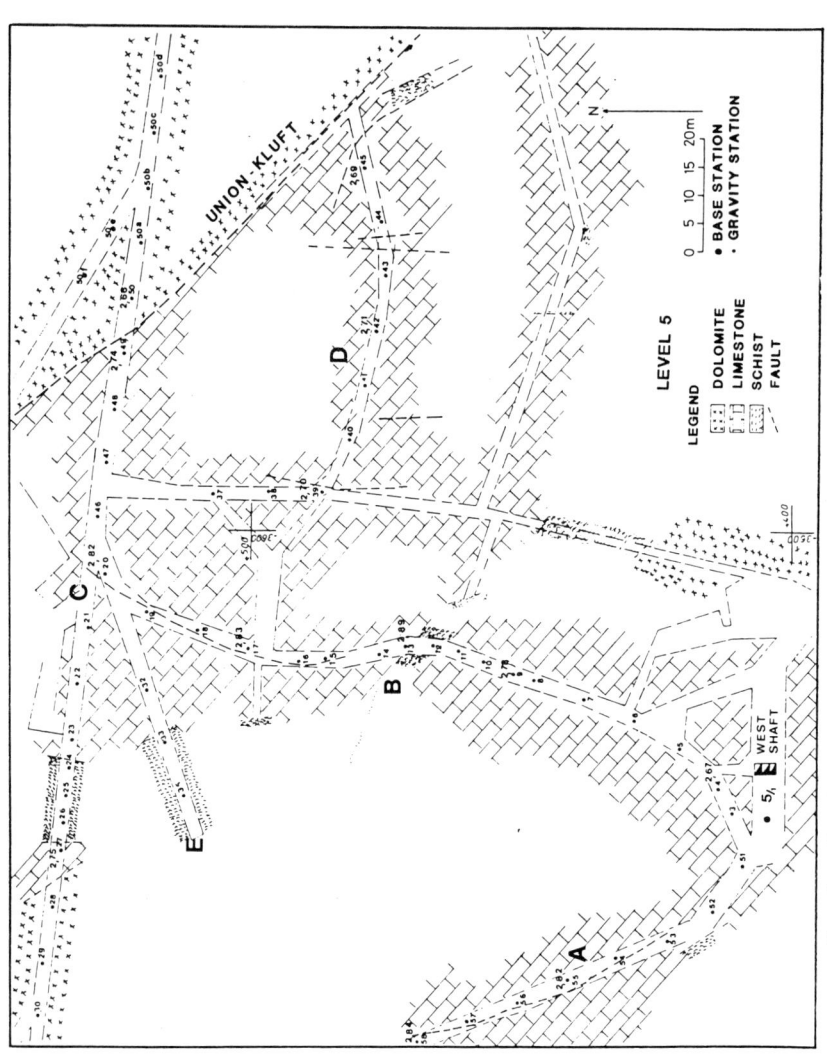

FIG. 8. Gravity station location and geology of the 5th level (251 m subsurface).

of the stations were established using optical levelling methods. Statistical analysis of the errors indicated that gravity data is accurate to approximately 0·04 mgal. All the readings on a level were reduced with respect to the shaft station on that level and plotted in the form of a contour map with a contour interval of 0·1 mgal (Fig. 9).

The gravity values 'regionally' decrease with the north-easterly trend. Superimposed on this field are the fluctuations related to areas which are either tectonically disturbed or where we found lithologic changes (limestone–dolomite–schist boundaries). Although the aerial extent of this survey is small and the regional gravity will be uncertain, a 'trend map' and a 'residual map' of this area were prepared just for a comparison with the Bouguer map. The trend map (Fig. 10) indicates a uniform N28°E decrease of gravity at the rate of 1·18 mgal/100 m. The observed gradient is much higher than the regional gravity gradient in this region which implies considerable local relief within the basement rocks. A slight distortion of the contours in the lower left corner is due to the lack of sufficient distribution of data points.

A first-order 'residual' gravity map of this level is presented in Fig. 11. The 'residual' surface is determined by subtracting the first-order fit from the Bouguer gravity surface and is a better approximation to the gravitational field produced by the local density variations. This map reveals some interesting features. Negative anomalies 'I' and 'IA' are associated with low density (2·70 g/cm^3) limestone which is faulted against the high density region of II ('D' in Fig. 8). The gravity anomalies 'IA' are quite interesting as the densities in this region are higher (2·80 g/cm^3) than the surrounding rocks but the observed anomalies are negative. This situation can be explained by the occurrence of the fault in gallery 'A' of Fig. 8, which separates, as in anomaly I, high density material II from region IA. The same anomaly pattern is observed in III where schist and limestone are faulted against dolomite. The gravity gradient is quite significant near Union Kluft and is associated with the dip of the fault above the level of measurements. The quantitative interpretation of this map was not attempted because of insufficient station distribution, but qualitatively it helps to understand the geological situation of this area.

3.1.2. Level 6

This level is 92 m below level 5. The gravity profiles, hand sample densities and the lithology along the galleries of this level are shown in Fig. 12. The rock type is mainly limestone without major structure occurrence around the galleries. The gravity profiles show only the trend decreasing towards

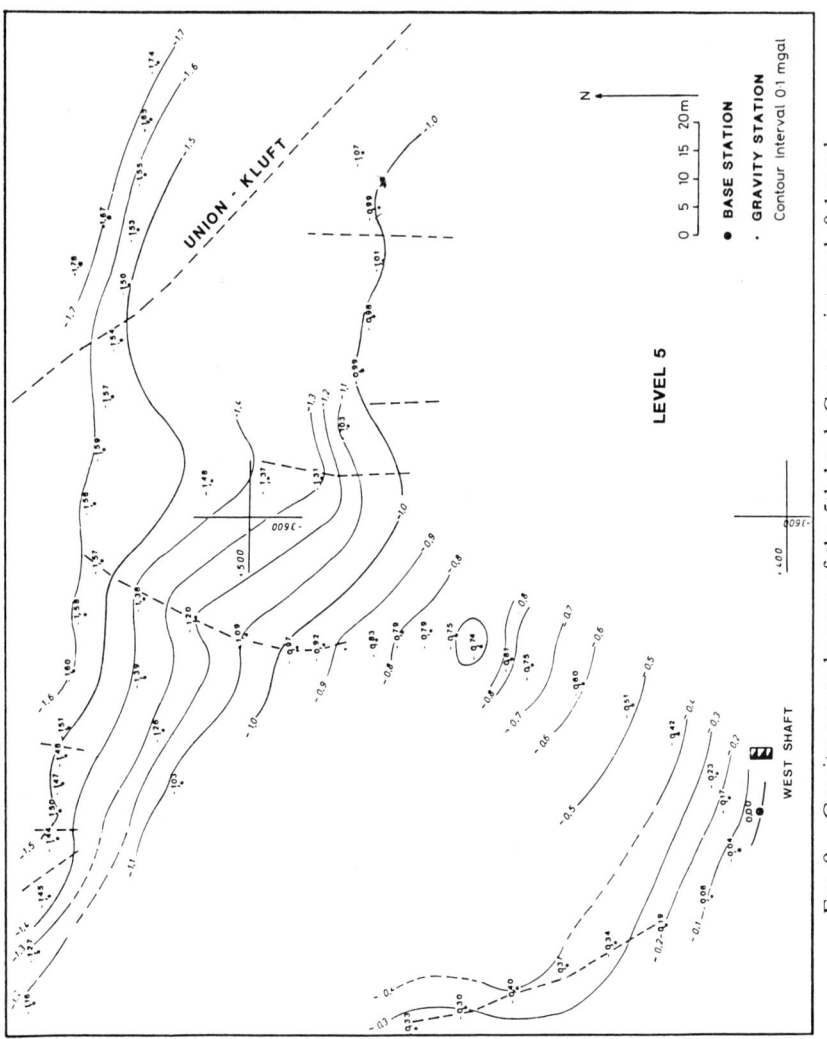

Fig. 9. Gravity anomaly map of the 5th level. Contour interval, 0·1 mgal.

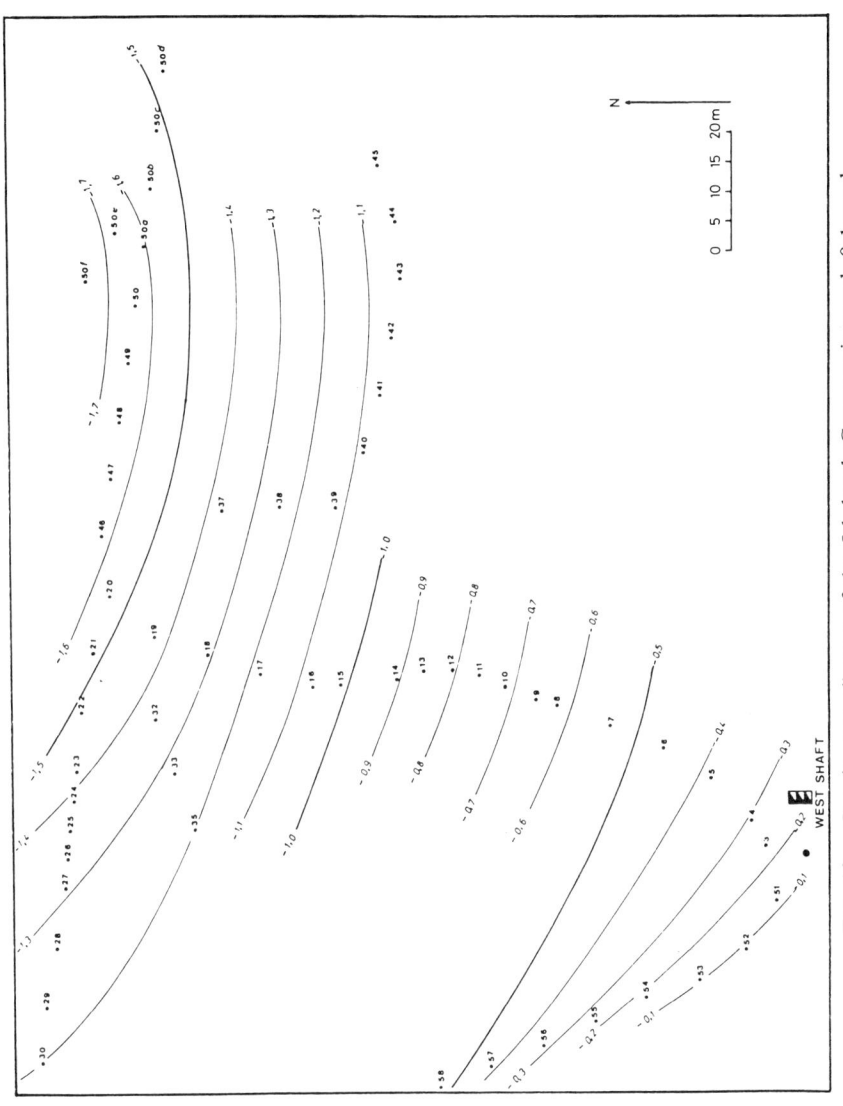

Fig. 10. Gravity 'trend' map of the 5th level. Contour interval, 0·1 mgal.

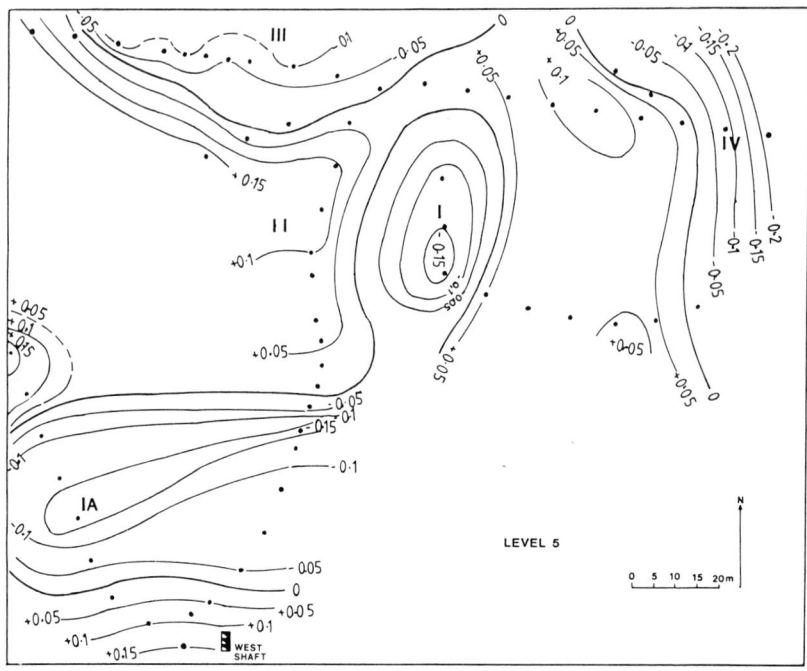

FIG. 11. 'Residual' gravity anomaly map of the 5th level. Contour interval, 0·05 mgal.

north, the same as observed on the 5th level, without any appreciable anomaly.

3.1.3. *Level 7*

The geology, station location and hand sample densities are shown in Fig. 13. All the gravity stations are located in limestone, but schist (2·66 g/cm^3) occurs to the left and above galleries A and B. The northwestern part of the map 'C' is partly mineralised. The average hand sample density is about 2·90 g/cm^3 in this part. This level is only 42 m below level 6 but a significant change in the gravity anomalies (Fig. 14) can be seen with respect to level 6. The change in trend is associated with the structural changes between the upper and lower levels. The positive anomaly along profile 'b' is associated with the low density schist occurring above level 7. Positive anomalies 'c_1' and 'c_2' in the upper left hand corner of the map coincide remarkably well with the known outlines of the weakly

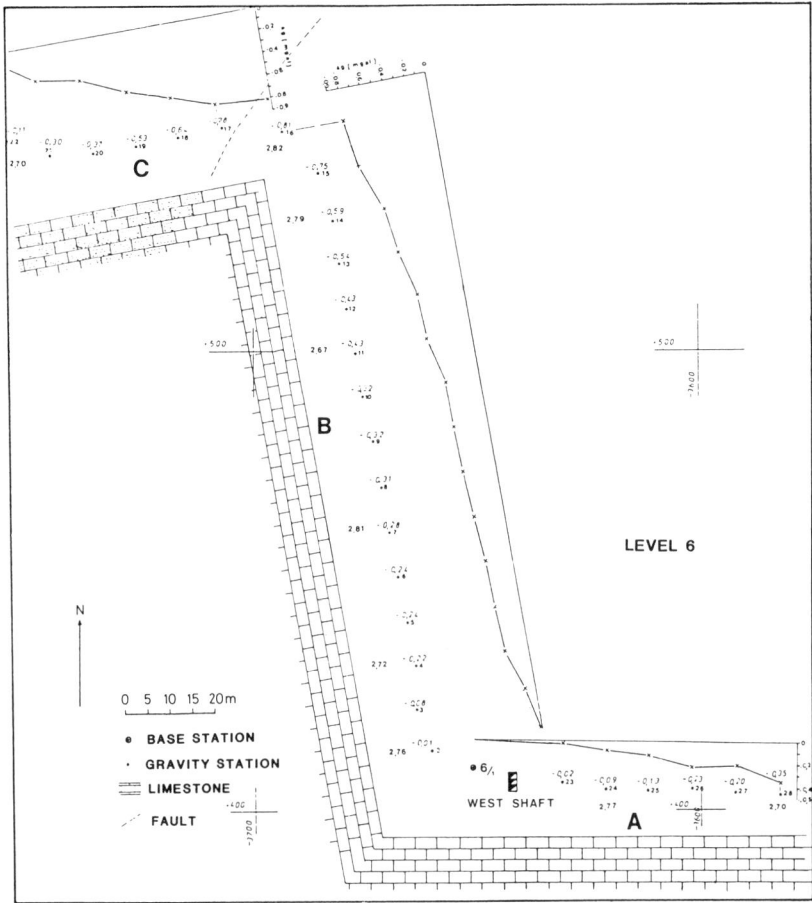

FIG. 12. Geological setting and gravity profiles of the 6th level.

mineralised limestones. Traces of lead mineralisation were found here but none approached ore grade. Profiles 'a' and 'd' represent only a trend.

3.1.4. Level 8

A compilation of the geology along the galleries at this level is presented in Fig. 15. The fault Union Kluft observed 185 m above this level (5th level) can also be seen on this level. A good distribution of the gravity stations at this level could be made because of sufficient available galleries, but all the galleries were not occupied because some of them were collapsed or

56　　　　　　　　　　　　A. HUSSAIN

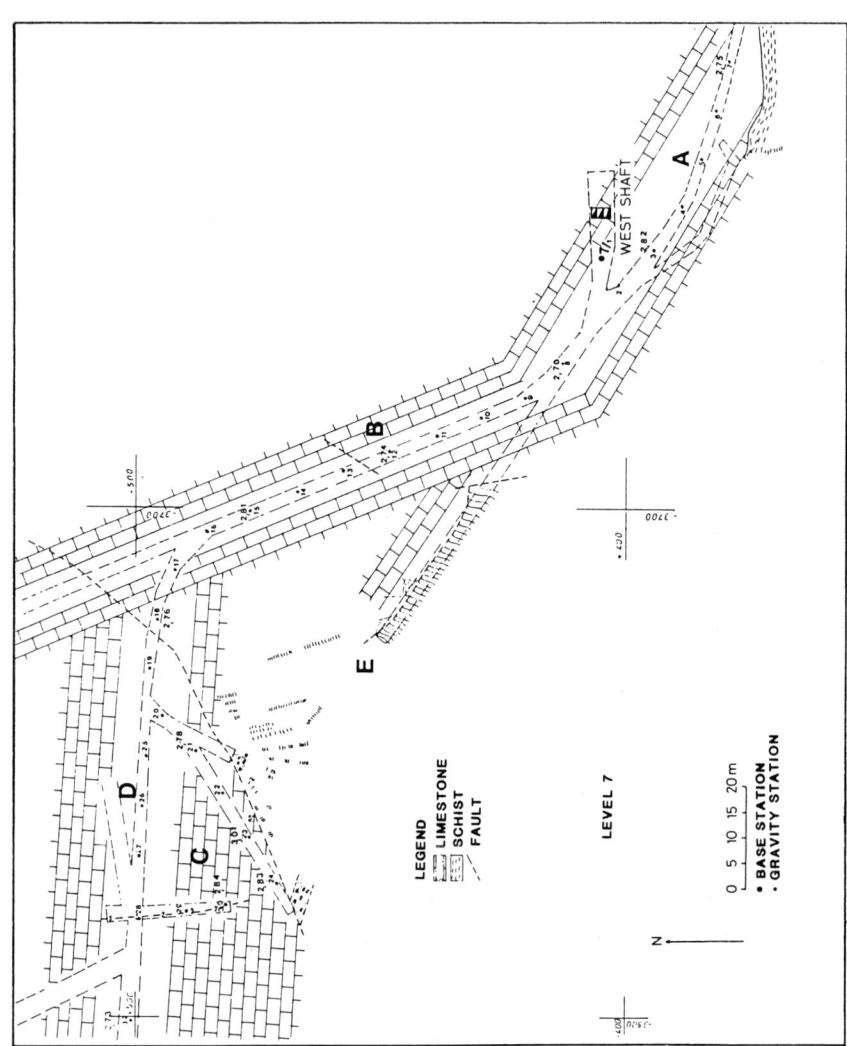

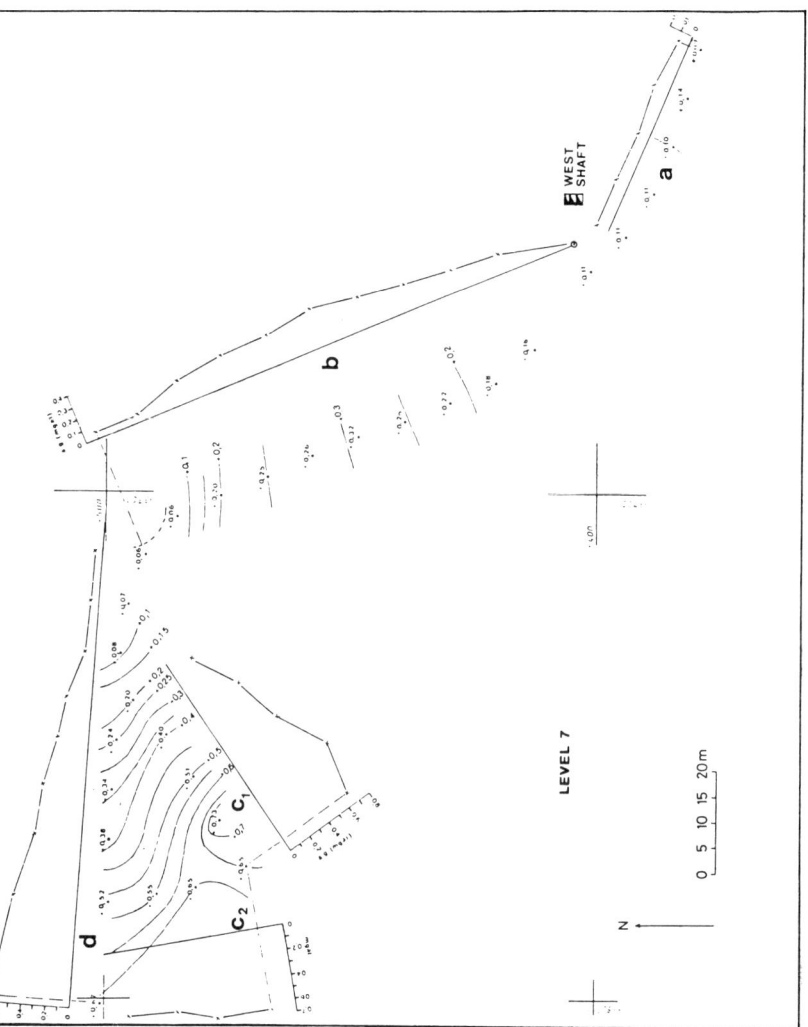

FIG. 14. Bouguer gravity map of the 7th level.

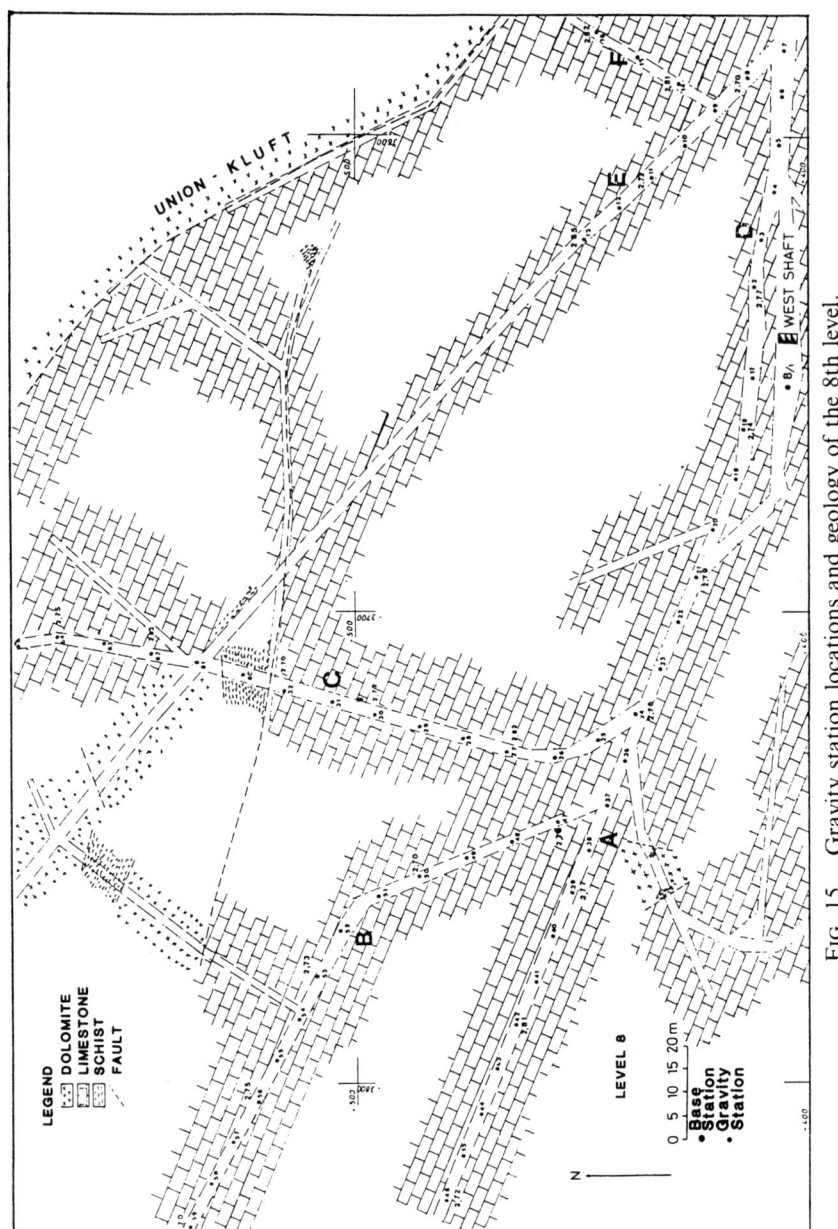

FIG. 15 Gravity station locations and geology of the 8th level.

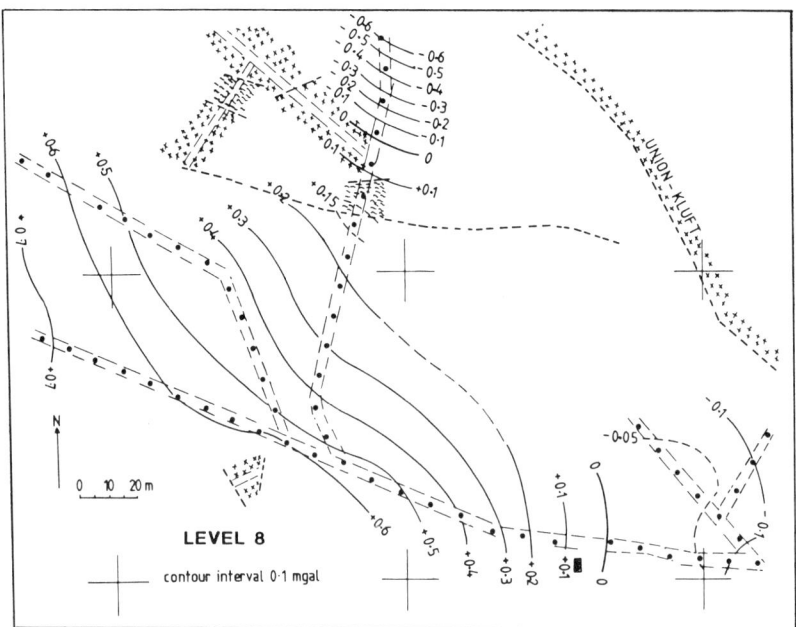

FIG. 16. Gravity map of the 8th level. Contour interval, 0·1 mgal.

already closed and some of them were full of water and measurements in these galleries were not possible.

The results of the gravimeter survey are shown in the form of gravity contours with a contour interval of 0·1 mgal in Fig. 16. This map can be compared with the corresponding map of level 5 (Fig. 9). Gravity trend is the same in both cases, i.e. north-easterly decreasing. Most of the contours express this trend. A small gravity anomaly is observed along gallery C which is associated with the presence of schist. The gravity gradient increases significantly as the Union Kluft is approached.

3.1.5. Stefani Shaft, Levels 11 and 12

After establishing the fact that the gravity method (with careful consideration of surrounding geology) can be applied as an exploration tool even in areas where structural features are rather small and the sulphide mineralisation of low grade, it was then decided to apply this method to a more practical problem.

In the eastern part of the Bleiberg mines the sulphide deposits are mainly of the vein type but the veins do not occur as continuous bodies because of tectonic disturbances. A massive vein with high galena concentration was mined-out from the 10th level (about 600 m subsurface) to the 12th level where it suddenly disappeared. Extensive test drilling was carried out to find the possible extent of the vein but with no success. It was decided to carry out a gravity survey in the galleries surrounding the vein. A total of 90 gravity stations were established in two stages. All the stations exist only in limestone and other rock types (dolomite or schist), or a major structure ore are not known to occur in the area of gravity coverage.

The average hand sample density from 71 samples collected from the 11th and 12th levels was determined as $2.72 \pm 0.04 \, \text{g/cm}^3$. The host rock

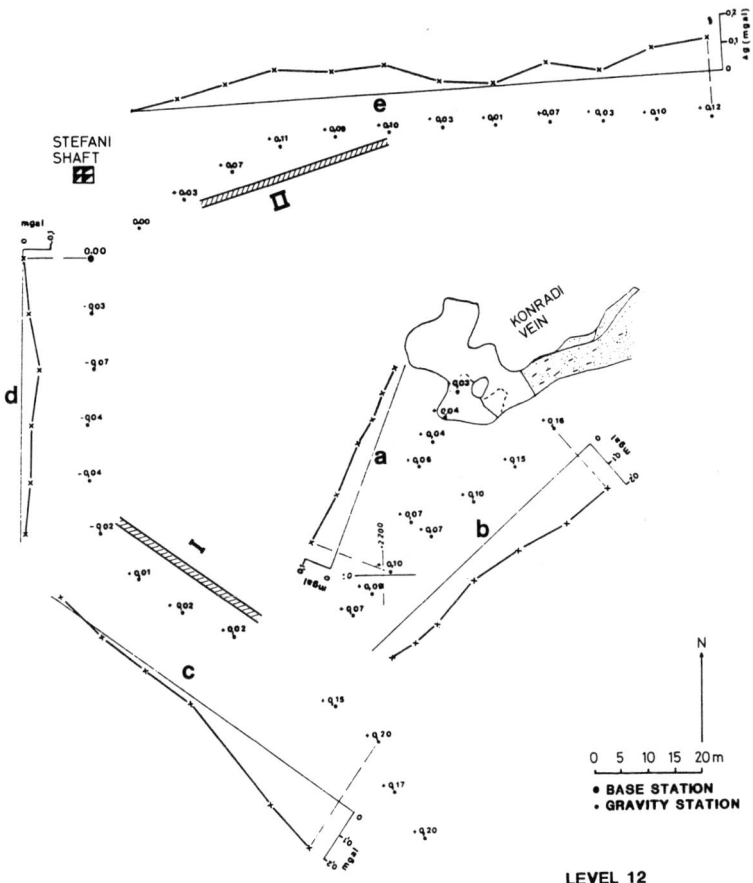

FIG. 17. Gravity profiles on the 12th level along galleries around the Stefani shaft.

seemed fairly uniform and the density contrast involved is substantially higher than in the West shaft area which provided an ideal target for gravity. Only the results of a part of the gravity survey surrounding the vein is of interest and will be discussed.

The station locations and the gravity profiles are shown in Fig. 17. Profiles 'a' and 'b' which lie near the mined-out vein show only a gravity trend without any appreciable anomaly. Profile 'c' is representing a sort of trend in its lower part (the south-easterly trend continues along the gallery not shown in the figure). However, a change in trend is observed from stations marked with shaded lines which seems to be a weak positive anomaly superimposed on the trend. This local anomaly (after subtracting from the trend) has a magnitude of only $+0.05$ mgal and can be neglected. However, it occurs at a favourable location with respect to the lost vein which will cross these stations at about 35–40 m in depth if its projection is made with the same dip as that observed at upper levels.

The most significant positive anomaly at this level was observed on profile 'e'. The gravity trend along the gallery is decreasing towards the west but a gradual increase, opposite to the trend, is observed in the shaded zone II which reaches a maximum of $+0.11$ mgal. Considering the surrounding geological setting, this anomaly was interpreted as caused by the occurrence of ore under this anomaly. Test drilling was suggested at locations I and II. A vein, about 1·5 m wide, with about 20 % lead concentration was found underneath anomaly I and only traces of lead were found underneath anomaly II.

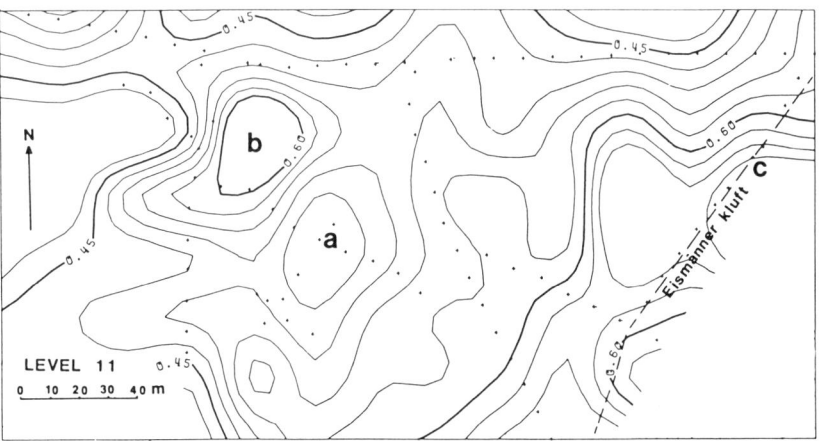

FIG. 18. Bouguer gravity map on the 11th level of the Stefani shaft.

At a later stage, another 100 gravity stations were established on the 11th level (51 m above level 12). The contoured gravity map with the station location is presented in Fig. 18. The anomalies clearly demonstrate the local gravity highs and lows. Anomalies 'a' and 'b' were interpreted as caused by the north-westerly dipping ore body. Negative anomaly 'a' is interpreted as an anomalous mass lying above the level of measurement and positive anomaly 'b' is possibly caused by the same dipping anomalous zone which at this point occurs below the level. Anomaly 'c' is caused by the presence of the fault the 'Eismänner Kluft' which is the boundary between limestone (left side of the fault) and dolomite.

4. CONCLUSION

Our experience in the Bleiberg mine establishes the gravity method as a useful underground exploration technique if employed with a recognition of its limitations and if full use is made of geological information in the interpretation of gravity data. The main advantage of such surveys is that observations of the gravitational field at different levels of the mine can be interpreted by two different techniques.

The gravity maps and profiles can be constructed and the trend of the contours and their disposition can be interpreted in terms of mass distribution laterally as well as vertically. Bulk densities can be determined from gravity measurements taken at stations lying vertically one above the other. An underground gravity survey has a definite advantage over surface work since the accuracy of an underground survey is greater due to the effect of topography being smoother at greater depth and, secondly, buried ore bodies are within the resolution of an underground survey. But the major difficulty in such a survey is that the gravity data is not sufficient to permit accurate interpretation. Therefore, quantitative interpretation of the Bouguer anomalies was not attempted in the given examples. Qualitatively it was possible to mark the high as well as low density zones.

ACKNOWLEDGEMENTS

The author wishes to express his indebtedness to the BBU Bleiberg and the Fohnsdorf Mining Authorities for permission to carry out measurements in the mines and their cooperation during the project. I am grateful to Dr G. Walach and Professor F. Weber for their help and advice. The editors of

Geophysical Prospecting and *Geoexploration* are thanked for permission to produce Figs 2, 3, 4 and 5 which came from earlier publications by the author.[5,13] Finally, I would like to express my gratitude to the 'Bunderministerium für Wissenschaft und Forschung' for their financial support.

REFERENCES

1. HAMMER, S. (1950) Density determination by underground gravity measurements. *Geophysics* **15**, 637–52.
2. SMITH, N. J. (1950) The case for gravity data from boreholes. *Geophysics* **15**, 605–36.
3. DOMZALSKI, W. (1955) Three-dimensional gravity survey. *Geophys. Prosp.* **3**, 1–55.
4. ALGERMISSEN, S. T. (1961) Underground and surface gravity survey, Leadwood, Missouri. *Geophysics* **26**, 158–68.
5. HUSSAIN, A., WALACH, G. and WEBER, F. (1981) Underground gravity survey in Alpine regions. *Geophys. Prosp.* **29**, 407–25.
6. MCCULLOH, T. H. (1965) A confirmation by gravity measurements of an underground density profile on core densities. *Geophysics* **30**, 1108–32.
7. MCCULLOH, T. H. (1967) Borehole gravimetry: new developments and applications. *Proc. 7th World Petrol. Congr.*, Vol. 2, Elsevier, London, pp. 735–44.
8. SUMNER, J. S. and SCHNEPFE, R. N. (1966) Underground gravity surveying at Bisbee, Arizona. *Mining geophysics I*, Soc. Explor. Geophys., Tulsa, pp. 243–51.
9. LAFEHR, T. R. (1981) Apparent density from borehole gravity surveys. Presented at the 51st SEG Meeting, Los Angeles.
10. HEARST, J. R. and MCKAGUE, H. L. (1976) Structure elucidation with borehole gravimetry. *Geophysics* **41**, 491–505.
11. HEARST, J. R. (1968) Terrain correction for borehole gravimetry. *Geophysics* **33**, 361–2.
12. HINZE, W. J., BRADLEY, J. W. and BROWN, A. R. (1978) Gravity survey in the Michigan basin deep boreholes. *J. Geophys. Res.* **83**, 5864–8.
13. HUSSAIN, A. and WALACH, G. (1980) Subsurface gravity measurements in a deep intra-Alpine Tertiary basin. *Geoexploration* **18**, 165–75.
14. HUSSAIN, A. (1979) Untertagegravimetrie in alpinen Gebieten mit besonderer Berücksichtigung des Blei-Zink-Bergbaus Bleiberg-Kreuth (Kärnten). Ph.D. thesis, Mining and Metallurgy University, Leobon.
15. KOSTELKA, L. (1971) Beiträge zur Geologie der Bleiberger Vererzung und deren Umgebung. *Carinthia* **II**, 63–75.
16. BECHSTÄDT, Th. (1978) Faziesanalyse permischer und triadischer Sedimente des Drauzuges als Hinweis auf eine grossräumige Lateralverschiebung innterhalb des Ostalpins. *Jahrbuch der Geologischen Bundesanstalt* **121**(1), 1–122.

Chapter 3

DIGITAL FILTERING OF GEOPHYSICAL LOGS

J. G. CONAWAY

Seismograph Service Corporation, Tulsa, Oklahoma, USA

SUMMARY

Raw geophysical logs are invariably less than ideal in terms of accuracy and spatial resolution. In general, the logs can be improved by some type of processing of the data. Many processing techniques are available to enhance the accuracy of geophysical logs and facilitate the interpretation in various ways. This review deals with one class of processing technique—linear digital filtering—applied to geophysical logs.

1. INTRODUCTION

Until the mid 1970s, the great majority of borehole logs were recorded directly in the form of analog traces using a chart recorder of some type. Commercial logging companies went to great lengths to perform corrections to the various logs in real time using analog electronic circuits in the logging trucks. If numerical calculations were needed in an attempt to quantify the interpretation, these analog traces were digitised manually or semi-automatically with an optical digitiser, a process which tends to be both inefficient and inaccurate.

Digital logging systems have now become fairly common. The numerous advantages of digital systems in terms of flexibility and relative immunity to noise are by this time well known. Of particular interest here is the fact that digital logs are amenable to computer processing. Cumbersome analog correction circuits can to a large extent be replaced by relatively simple computer algorithms, which can easily be changed in case the system is modified or redesigned. Additional processing techniques are available (for either on-line or post-processing application) which go far beyond what can practically be accomplished with analog systems.

This chapter presents a review of the state of the art in one area of digital processing of geophysical logs—the application of digital filtering techniques. Section 2 presents a review of some digital time series analysis techniques which have been applied to geophysical logs. Succeeding sections review the existing literature on specific applications. An earlier review was given by Lindseth.[1]

2. DIGITAL FILTERING

Filtering is a hardware or software operation performed on a function or waveform (the input function) with the goal of producing an output function or waveform having frequency characteristics altered in some desired way. Generally speaking if the effect of the filter is independent of the amplitude and polarity of the input waveform then the operation is said to be linear. Historically, filtering of data (including geophysical data) has generally been performed in an analog or continuous manner by electronic circuits. Over the past few decades rapid advances have been made in the development of digital data recording and processing techniques, including digital filtering. Digital filters can produce results essentially identical to their analog counterparts, and, in addition, for many applications offer a great deal more simplicity and versatility than analog filters. In ultra-high speed applications hardware analog filters still hold the edge, although that advantage narrows each year.

The mathematical operation describing the process of applying a linear analog filter $f(t)$ to an input signal $i(t)$ is the convolution integral:

$$h(t) = \int_{-\infty}^{\infty} f(\tau) i(t - \tau) \, d\tau \qquad (1)$$

where $h(t)$ is the output function and τ is the convolution lag term. A convenient short form representation of the convolution operation is

$$h(t) = f(t) * i(t) \qquad (2)$$

A treatment of convolution can be found in most text books dealing with Fourier analysis or time series analysis.[2,3]

The function $f(t)$ that characterises a linear filter is called its impulse response. Knowing the response of a linear system to an impulse (event of infinitesimal duration in time or space) allows the response of the system to any other input function to be determined, as given by eqns (1) and (2). A digital filter is a series of discrete numerical coefficients or weights. These weights represent the digitised impulse response of the filter, i.e. the response of the filter to an impulse input. The term discrete or digital convolution denotes a simple arithmetic procedure involving multiplication and addition which is used to apply a digital filter to a time series (digital data gathered at equal intervals of time or distance; in our case, a digital log or digitised analog log). To illustrate digital convolution, assume that nine discrete log data measurements have been made at equal intervals along a borehole. This is the raw log R, represented by the symbols

$$R = R_1, R_2, R_3, \ldots, R_9$$

Assume also that a three-coefficient digital filter $F = F_1, F_2, F_3$ is to be applied to the raw log to produce a processed log $P = P_1, P_2, P_3, \ldots$ As shown in eqn (2) this procedure is expressed symbolically as $R * F = P$.

The diagram in Fig. 1 shows the procedure. First, the filter is reversed and positioned such that the first coefficient of the filter is registered with the first data point in the raw log. Each value of the raw log is multiplied by the filter coefficient with which it is registered, and the results of these separate multiplications are summed to produce the first value of the processed log. Notice that coefficients F_2 and F_3 are multiplied by zero, since they are not registered with any value of the raw log. In order to

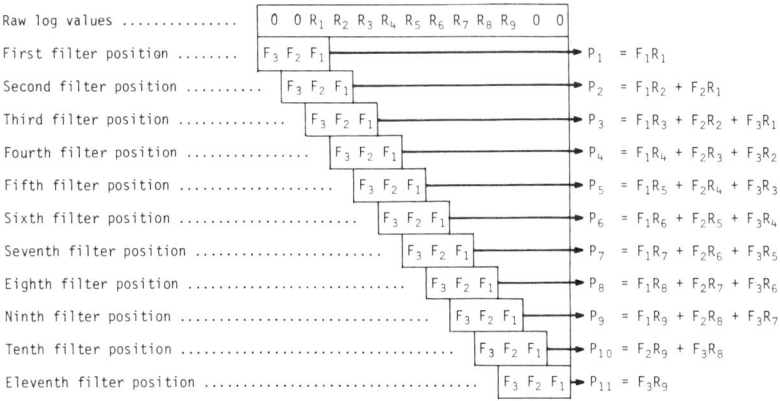

FIG. 1. Diagram illustrating the digital convolution procedure, as explained in Section 2.

obtain the second value in the processed log, the digital filter is shifted one unit to the right and the multiplication–addition process is repeated. This continues as shown in Fig. 1 until the entire raw log has been processed.

Digital convolution is commonly used to apply simple smoothing or low-pass filters (so called because only the lower frequency components of the input function are allowed to pass by the filter) and other more sophisticated filters designed to discriminate against coherent noise on the basis of frequency content. In addition, digital filters can be used for deconvolving an input function, that is, removing the unwanted effect of a previous filtering operation. This unwanted effect may represent the distorting influence of a sensor and/or associated instrumentation, or it may represent the distorting effect inherent in the physical phenomenon being measured. In other words, if the desired input function can be regarded as having been convolved with a distorting filter (by nature or by the instrumentation) then it is often possible to deconvolve the distorted signal to obtain the desired undistorted input function, within limits imposed by noise.

Figure 2 illustrates deconvolution and smoothing from a frequency domain perspective. Figure 2(a) shows the noise free amplitude spectrum of an impulse (i.e. an event of infinitesimal duration). This spectrum is said to be white, i.e. all frequencies represented equally. Figure 2(b) shows the noise-free impulse response distorted by nature or by the sensing system. The process of deconvolving this distorted signal requires that the attenuated high frequency information be restored. When random noise is present in the signal to be deconvolved, the amplitude spectrum of the system impulse response is as shown in Fig. 2(c). This may be smoothed by applying a low-pass filter to attenuate high frequency noise and improve the signal-to-noise ratio. After deconvolution of the noisy signal (Fig. 2(c)) the signal has been restored to the white spectrum of an impulse, but the noise has also been amplified by a corresponding amount (Fig. 2(d)), and may dominate the signal completely. This may necessitate using a low-pass filter to reduce the amplified higher frequency noise. Often the best results will be obtained by some combination of smoothing and deconvolution to sharpen the signal and still keep noise to a tolerable level.

If the phenomenon being measured is a continuous function, then obviously it must be digitised or sampled before it can be operated upon by digital filters. The choice of the digitising interval is important. As an example, consider what happens when the simple sinusoidal waveform shown in Fig. 3(a) is digitised. As the digitisation interval increases (Fig. 3(b), (c) and (d)) the shape of the waveform becomes distorted until

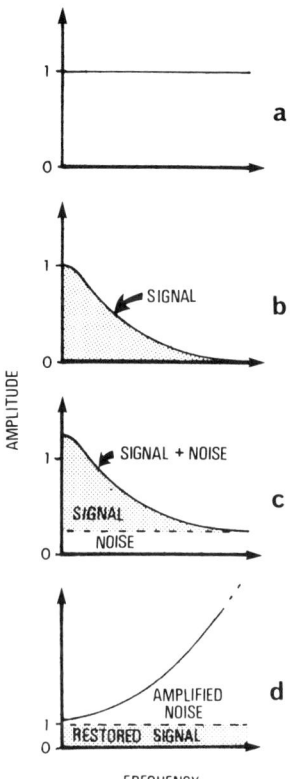

FIG. 2. Diagrammatic representation of amplitude spectra showing the effect of deconvolution on the recorded log. (a) Flat or white spectrum of an impulse. (b) Spectrum of an idealised noise-free logging system impulse response (degraded by the logging system or natural effects). Application of the appropriate inverse filter would ideally recover the flat spectrum of the impulse shown in (a). (c) The system impulse response shown in (b) with random (white) noise added. (d) Application of the inverse filter to the noisy data in (c) will recover the flat spectrum of the impulse, but with the correspondingly amplified noise inextricably added in.

finally (Fig. 3(e)) the frequency characteristics of the original waveform are completely changed. This effect is known as aliasing, when a high-frequency component of a signal appears in the digitised form to be of a lower frequency than it actually is, because the digitisation interval was too large.

Aliasing can seriously degrade the quality of the digitised log, as shown

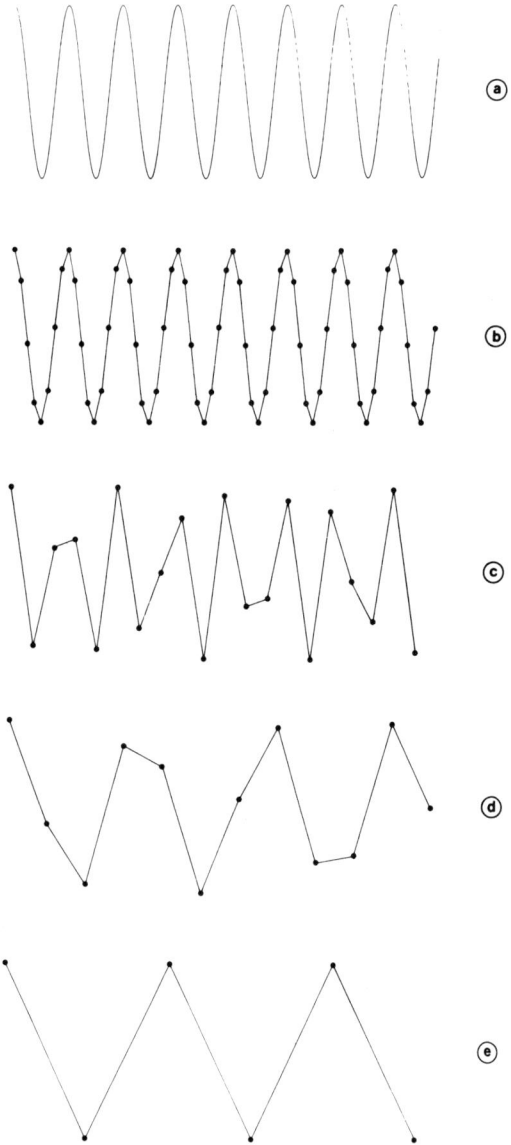

FIG. 3. The pure sinusoid shown in (a) can be sampled so as to maintain frequency content *and* appearance (b). As sampling frequency decreases as shown in (c) and (d), appearance degrades until (e), where the apparent frequency of the sampled sine wave is half the true frequency.

in Fig. 4. Here a simulated analog log (Fig. 4(a)) is digitised at two different intervals showing the loss of depth resolution with increasing digitising interval. The sample interval must be chosen such that the shape of the log is not distorted appreciably. To avoid aliasing, the input function should be sampled such that at least two samples per cycle are obtained for the highest frequency component present in the signal. If that sampling rate is impractically high then aliasing can be avoided by attenuating the higher frequency components with an analog (e.g. electronic) low-pass filter before sampling.

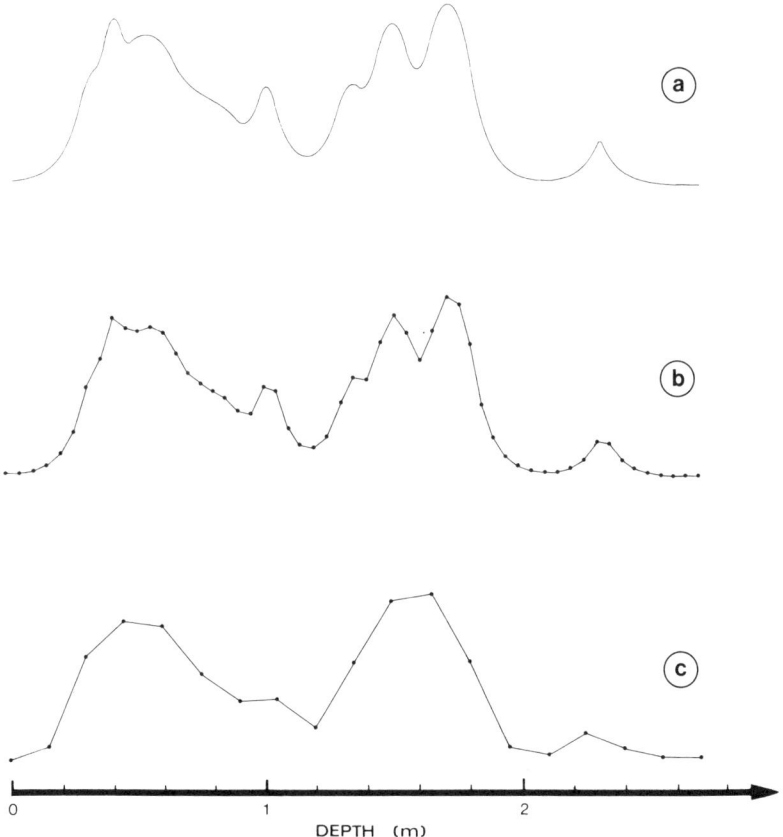

FIG. 4. A noise-free synthetic log (a) is sampled at two different frequencies, (b) and (c), illustrating the loss of information as the sampling frequency decreases. Application of an inverse filter cannot restore information lost by undersampling.

2.1. Frequency Domain Methods

Fourier transforms are useful in linear filter design. For continuous functions, the input function $i(t)$ is transformed from the 'time domain' to the 'frequency domain' by integrating as follows:

$$I(\omega) = \int_{-\infty}^{\infty} i(t) \exp(-j\omega t)\, dt \qquad (3)$$

where $I(\omega)$ is the Fourier transform of $i(t)$, and $j = \sqrt{-1}$. $I(\omega)$ is generally complex (i.e. contains real and imaginary parts) and expresses the frequency content of $i(t)$ and the corresponding phase relationships.[2,3]

Given the Fourier transform $D(\omega)$ of the desired output function $d(t)$, the transform $F(\omega)$ of the time domain inverse filter $f(t)$ is found by complex division

$$F(\omega) = \frac{D(\omega)}{I(\omega)} \qquad (4)$$

Equation (4) is made possible by the fact that multiplication of the Fourier transforms of two time domain functions is equivalent to the convolution of those functions in the time domain. The desired time domain filter is found by applying the inverse Fourier transform to transform $F(\omega)$ back to the time domain:

$$f(t) = \frac{1}{2\pi} \int_{-\infty}^{\infty} F(\omega) \exp(j\omega t)\, d\omega \qquad (5)$$

In the case of digital functions, discrete Fourier transforms may be used on a computer to perform the required forward and reverse transform operations.[4-6] Using the fact (once again) that multiplication in the frequency domain is equivalent to convolution in the time domain, the filter may also be applied in the frequency domain. The procedure is to transform the digital raw log (or other input) to the frequency domain, apply the frequency domain filter by complex multiplication (a FORTRAN library routine on many computers) and transform back to the time domain to obtain the filtered log. In some cases this may be a more efficient technique computationally since digital convolution of long time series is a time-consuming operation.

2.2. Exact Digital Filters

Given a digital input function $I_0, I_1, I_2, \ldots, I_n$ which is the system impulse response, a digital inverse filter may be derived which will convert that input function into a desired output function $D_0, D_1, D_2, \ldots, D_p$ using Z-transform techniques.[7,8] Assuming I_0 occurs at time zero, the input Z-transform of the function may be written as

$$I(Z) = I_0 + I_1 Z^{-1} + I_2 Z^{-2} + \cdots + I_n Z^{-n} \qquad (6)$$

Values of I to the left of time zero would have been written with positive exponents of Z; some authors use the reverse of this convention. Equation (6) is seen to be a polynomial in Z with constant coefficients. Similarly, the desired output may be written as

$$D(Z) = D_0 + D_1 Z^{-1} + D_2 Z^{-2} + \cdots + D_p Z^{-p} \qquad (7)$$

What we want to find is a filter $F(Z)$ such that polynomial multiplication with $I(Z)$ gives $D(Z)$:

$$D(Z) = F(Z) \cdot I(Z) \qquad (8)$$

Solving for $F(Z)$:

$$F(Z) = \frac{D(Z)}{I(Z)} \qquad (9)$$

or, in the case of pure deconvolution where the desired output is a unit impulse $D(Z) = 1$, the inverse filter is of course given by

$$F(Z) = \frac{1}{I(Z)} \qquad (10)$$

Equations (9) and (10) indicate that exact inverse filters for digital data may be obtained by a simple polynomial division of the input function into the desired output. In practice a slightly more involved procedure is required in some cases to avoid instability (endlessly increasing filter coefficients); this is explained clearly by Treitel and Robinson.[7]

Often, the exact inverse filter is infinitely long and must be truncated before use. This introduces some error into the output. In this case, another procedure such as optimum filtering may be preferred; this approach will be discussed in Section 2.3.

An extension of the Z-transform approach leads to the computation of recursive filters.[9] These filters operate on new data and also previously processed data as they move along the input function. Using this technique, results can often be achieved with a short recursive filter which might otherwise require a very long (and therefore computationally inefficient) standard convolution filter.

2.3. Optimum Filters

The least square error (LSE) or Wiener filter is one of a class of optimum linear filters. These filters are called optimum because the effect of the filter on the input function is optimised according to some criterion. The LSE inverse filter is a general filter which is designed to convert a given input function as nearly as possible into a specified desired output function, such that the sum of the squared errors or differences between the actual output and the desired output is a minimum. In geophysical logging applications, if an inverse filter is to be developed the input function is generally the system impulse response. The desired output may be an impulse, or it may be some other function as will be discussed later in this section.

The theory of the digital least square error filter is given in many places.[10,11] The LSE filter

$$(F_0, F_1, F_2, \ldots, F_m) \qquad (11)$$

is designed to convert a known input function or waveform

$$(I_0, I_1, I_2, \ldots, I_n) \qquad (12)$$

into a desired output function

$$(D_0, D_1, D_2, \ldots, D_p) \qquad (13)$$

as nearly as possible. In general the actual output from the filter

$$(A_0, A_1, A_2, \ldots, A_{m+n}) \qquad (14)$$

will not be the same as the desired output. As the name implies, the LSE filter is designed to minimise the sum of the squared errors, E, between the actual and desired output functions, where

$$E = \sum_{i=0}^{\infty} (D_i - A_i)^2 \qquad (15)$$

Since the inverse filter is to be applied using discrete convolution, it is known that

$$A_i = \sum_{j=0}^{m} F_j I_{i-j} \qquad \text{for } i = 0, 1, 2, \ldots, m+n$$

and

$$A_i = 0 \qquad \text{for } i > m+n \qquad (16)$$

Thus

$$E = \sum_{i=0}^{m+n}\left(D_i - \sum_{j=0}^{m}F_j I_{i-j}\right)^2 + \sum_{i=m+n+1}^{\infty} D_i^2 \qquad (17)$$

The sum of the squared errors, E, may be minimised by setting its partial derivative with respect to each of the filter coefficients equal to zero, giving

$$\frac{\partial E}{\partial F_k} = \sum_{i=0}^{m+n} 2\left(D_i - \sum_{j=0}^{m} F_j I_{i-j}\right)(-I_{i-k}) = 0 \qquad \text{for } k = 0, 1, \ldots, m \qquad (18)$$

or

$$\sum_{j=0}^{m} F_j \left(\sum_{i=0}^{m+n} I_{i-j} I_{i-k}\right) = \sum_{i=0}^{m+n} D_i I_{i-k} \qquad \text{for } k = 0, 1, \ldots, m \qquad (19)$$

The simultaneous solution of this set of normal equations (eqn (19)) gives the LSE filter. Canned computer programs[5] are available to perform these operations.

As stated above, if the system impulse response is the input function, the desired output may be simply an impulse. This will provide the maximum fidelity in restoring the flat frequency spectrum of the impulse, but also will amplify noise as discussed earlier. A low-pass (smoothing) filter may be required to reduce this amplified noise. Thus it can be seen that in processing the data a two-step operation is involved. First, application of the inverse filter, followed by application of a low-pass filter. This may be reduced to a single step by using the low-pass filter coefficients as the desired output function. If this is done, however, a new LSE filter must be generated each time the smoothing filter is changed to meet altered noise characteristics.

2.4. Noise Rejection

In the process of filtering geophysical logs the goal is generally to enhance selected information-bearing aspects of the data while discriminating against undesired content or noise. The extent to which signal can be enhanced and noise rejected depends on the frequency spectrum of the signal, the noise and the desired output.

In many applications of time series analysis, sophisticated filters are

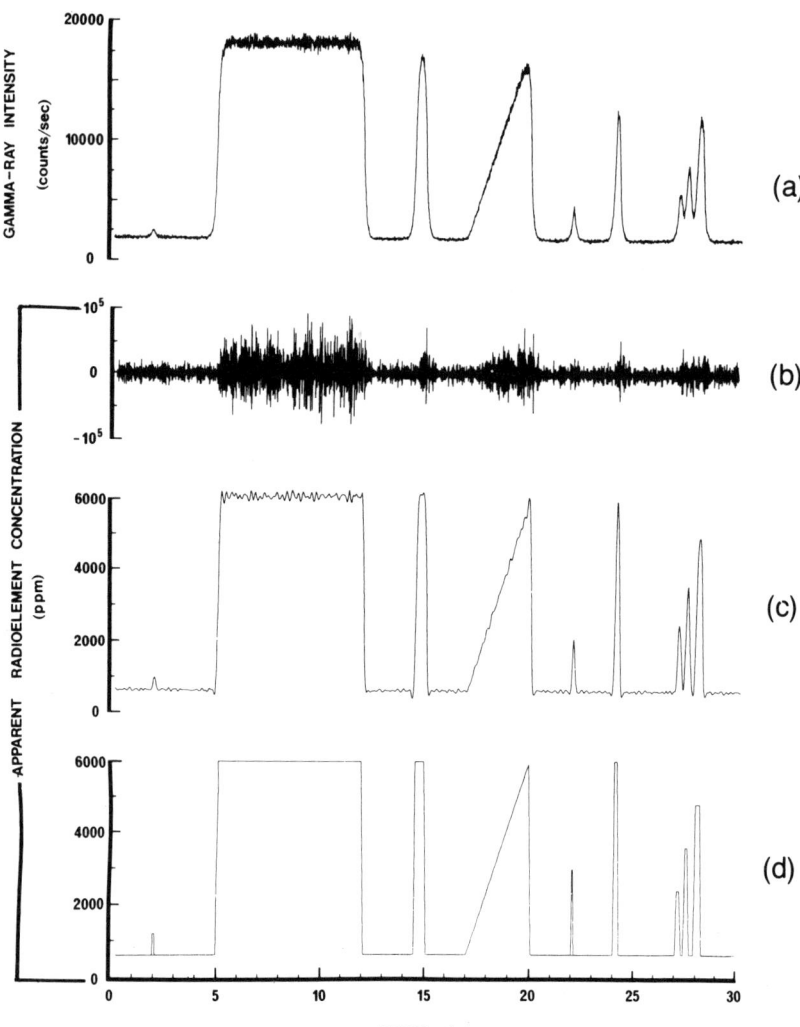

FIG. 5. Synthetic gamma-ray logs showing the effects of deconvolution and smoothing. (a) Computer generated digital log with a sampling interval of 1 cm, including pseudo-random simulated counting noise. (b) Deconvolved log after application of a 3-coefficient inverse filter (described in Section 7) to the data in (a). (c) Same deconvolved data as (b), smoothed with a simple low-pass filter. (d) Ideal noise-free profile corresponding to (c), for comparison.[12]

designed to discriminate against coherent noise on the basis of frequency characteristics. For many logging applications, however, a simple low-pass filter can do much to improve the appearance of a raw or previously processed log.

It is not necessarily intuitively obvious why high frequency content should be restored to a signal using deconvolution, and then removed again using a low-pass filter. The answer lies in exactly which frequencies are amplified and which are attenuated. The goal is to improve the frequency content as much as practical within limits imposed by noise. An example from Conaway and Killeen[12] illustrates this point with a computer-simulated gamma-ray log in a high activity radioactive sequence (e.g. uranium deposit). Figure 5(d) shows the simulated profile of radioelement concentration (ppm). Figure 5(a) is the synthetic digital gamma-ray log obtained by first convolving Fig. 5(d) with an appropriate impulse-response function, and then adding pseudo-random counting noise as determined by the count rate* (see Conaway and Killeen[12] for more details; gamma-ray log inversion will also be discussed in Section 7 of this review). The sampling interval in this log is 1 cm. Application of a simple deconvolution filter (described in Section 7) to Fig. 5(a) produces the extremely noisy profile of Fig. 5(b). A simple low-pass filter applied to these noisy data (Fig. 5(b)) yields Fig. 5(c), which can be compared with the original radioelement concentration profile, Fig. 5(d). It can be seen that the radioelement concentrations estimated in Fig. 5(c) are much improved over the raw log (Fig. 5(a)), while noise is still adequately controlled.

The ideal low-pass filter is the sinc function, a portion of which is illustrated in Fig. 6(a). This function $f(t)$ has the form

$$f(t) = \frac{\sin \omega t}{\omega t}$$

and ideally has a perfect cutoff in the frequency domain (assuming that the entire infinitely-long time-domain function is used). On the other hand, this filter causes ringing and overshoot (known as Gibb's phenomenon)[2] wherever a sharp transition is found on the raw log to which it is being applied. A simple low-pass filter which is entirely adequate for many logging applications is a full cycle cosine bell (Fig. 6(b)) with all coefficients being positive and normalised so that their sum is one. The exponential filter

---

* Radioactive decay processes are completely random. For count rates over about 10 counts per period the noise (uncertainty) is essentially Gaussian with standard deviation equal to the square root of the mean count rate.

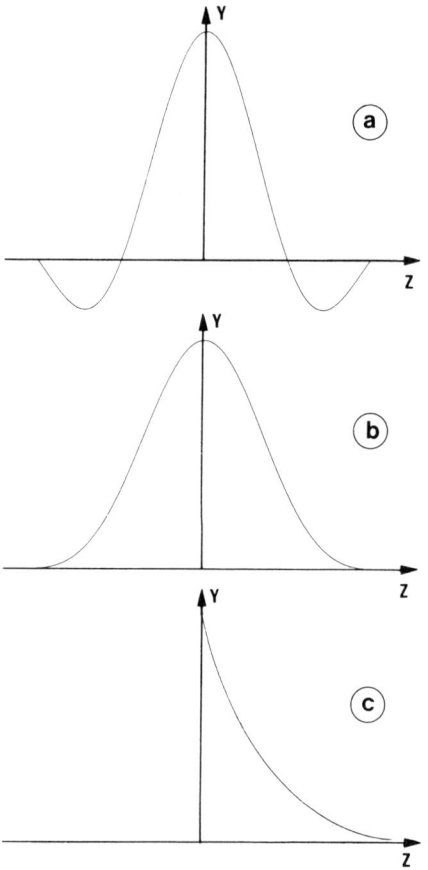

FIG. 6. Impulse responses of three different types of low-pass filters. (a) Sinc function, (b) cosine bell, and (c) exponential analog ratemeter.

generally used in analog ratemeter circuits (Fig. 6(c)) is one-sided, smearing the log in the direction of probe motion (see Sections 5 and 6).

The commonly used rectangular filter (equally weighted running mean) is not a particularly good low-pass filter, although it is adequate for some applications. Shultz and Thadani[13] compared several low-pass filtering techniques for reducing statistical variations in nuclear logs. They considered the problem of maintaining sharp discontinuities which are 'real' (e.g. bed boundaries) while discriminating against random noise. Linear techniques are limited in their ability to do this, and the authors concluded that non-linear techniques offer some advantages in certain situations.

3. ACOUSTIC LOGS

Foster et al.[14] applied digital filters to acoustic logs to change the effective transmitter–detector spacing. The conceptual model is illustrated in Fig. 7, which shows 3 dual detector acoustic probes having the same spacing from the transmitter to the first detector, but different spacings between the detectors. Generally in a dual detector probe the travel time to the near detector is subtracted from the travel time to the far detector to give (ideally) the interval transit time over a distance equal to the separation between the two detectors. In Fig. 7(a) the detectors are situated a small distance Δz apart where Δz is equal to the sample interval along the borehole, for convenience. In Fig. 7(b) the detectors are located a distance $N \Delta z$ apart, where N is an integer, and in Fig. 7(c) the detectors are separated by $M \Delta z$, where M is also an integer. Let L_a, L_b and L_c be the logs produced by probes a, b and c, respectively, in a given borehole. In the

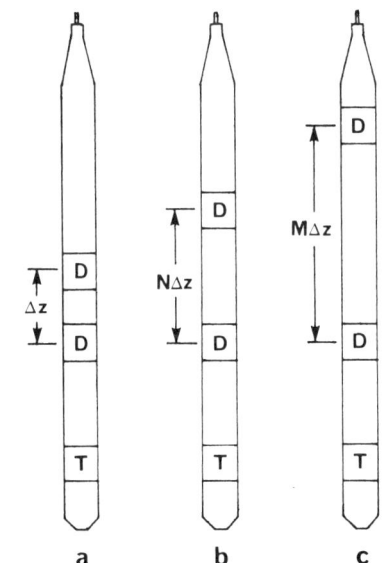

FIG. 7. Diagram of 3 dual detector acoustic probes with different spacings between the detectors, as explained in Section 3.

absence of noise, L_c will be essentially the same as $L_a * F_{ac}$ where F_{ac} is an M-coefficient digital filter, each coefficient having value $1/M$. Similarly, L_b will be essentially the same as $L_a * F_{ab}$ where F_{ab} is an N-coefficient digital filter, each coefficient having value $1/N$.

It is clear that a given log produced by probe c can be deconvolved to simulate a corresponding log from probe a if a suitable inverse filter F_{ca} can be developed. This inverse filter can be defined by the expression

$$F_{ca} * F_{ac} = 1$$

and the above-described operation denoted by

$$L_c * F_{ca} \simeq L_a$$

Finally, a log from probe c can be filtered to simulate a log from probe b as follows:

$$L_c * F_{ca} * F_{ab} \simeq L_b$$

where F_{ab} is an inverse filter defined by

$$F_{ab} * F_{ba} = 1$$

Ideally, then, an acoustic log from a probe of a given detector spacing can be processed to simulate another spacing if appropriate inverse filters can be developed as described above.

The function represented by F_{ab} and F_{ac} is a rectangular waveform (boxcar function) of length equal to the spacing between detectors. The exact digital inverse filter for this function is infinitely long and thus of no direct use. Having established this fact, Foster et al.[14] proceeded to derive least square error inverse filters for this application. As described earlier, these filters are optimum in that for a filter of a given length, the difference between the desired output and the actual output is minimised in a least squares sense.

An example from Foster et al.[14] is shown redrawn in Fig. 8, as an example. Figure 8(a) shows a portion of an acoustic log from a probe with a 3·1 m spacing between the detectors. A least square error inverse filter was applied to this log to produce a 'synthetic' log with an apparent detector spacing of 15 cm (Fig. 8(b)). This, in turn, is convolved with a rectangular function of length 90 cm to give a synthetic log with an apparent detector spacing of 90 cm (Fig. 8(c)). This can be compared with Fig. 8(d), a log obtained with an actual detector spacing of 85 cm. The two logs are quite similar, with differences between the logs not much greater than what would be expected between 2 runs with the same tool.

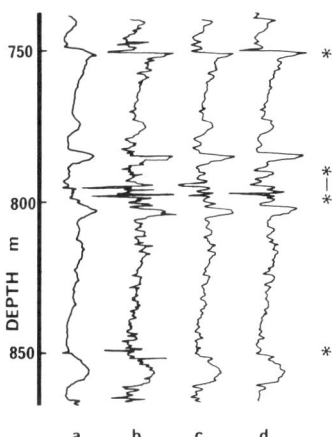

FIG. 8. Four acoustic logs redrawn from Foster et al.:[14] (a) 3·1 m spacing field log; (b) 0·15 m spacing log obtained by filtering log (a); (c) 0·9 m spacing log obtained by filtering log (b); (d) 0·85 m spacing field log. Asterisks mark noise events.

Foster et al.[14] point out that noise spikes on an acoustic log (due for instance to instrument microphonics or cycle skipping) will stand out more clearly on a long spacing log due to the inherent lack of sharp (high frequency) events on the latter. Such noise spikes are relatively easily recognised and removed by computer on a long spacing log compared to a short spacing log. Thus, the authors suggest that a long spacing log can be run, any noise spikes removed, and the log processed to produce a synthetic log of shorter effective spacing and correspondingly improved spatial resolution. These techniques give the option of postprocessing the log data to suit the requirements or preferences of the log analyst.

In a later paper, Runge and Powell[15] applied techniques similar to those used by Foster et al.[14] to alter the apparent spacing of acoustic logs. They presented many examples of both increasing the apparent spacing and decreasing the apparent spacing using digital filters.[15] The authors also discussed the effect of sampling interval on the quality of the processed log.

Modern acoustic logging probes generally have more transmitters and detectors to compensate for borehole irregularities and other effects. The basic idea presented by Foster et al.[14] is still valid; the effective vertical resolution of the log can be altered at least qualitatively by digital filtering techniques.

Certain other types of logging devices which have both a source and a detector can be filtered to alter the effective source–receiver spacing. In particular, sidewalled active nuclear tools such as gamma–gamma density should be amenable to processing by techniques similar to those presented by Foster et al.[14] The spatial resolution of these nuclear logs is generally better than that of acoustic logs, but in some applications, improved resolution may be desired. The signal obtained from a sidewalled nuclear probe is not a linear function of formation properties, however, therefore use of a linear inversion technique will be of qualitative use only.

4. RESISTIVITY LOGS

George[16] and George et al.[17] investigated the application of digital filters to resistivity logs. These authors made a number of simplifying assumptions to reduce the problem to a linear one. The spatial impulse response, $g(z)$, of a two-coil induction system with coil spacing L was regarded as fixed (independent of borehole and formation parameters), and given by:

$$g(z) = \frac{1}{2L} \quad \text{for} \quad -\frac{L}{2} \le z \le \frac{L}{2}$$

$$g(z) = \frac{L}{8z^2} \quad \text{for} \quad |z| > \frac{L}{2} \tag{20}$$

based on the geometric factor approach described by Doll,[18a] in which propagation and skin effects were neglected. The above system response function is illustrated in Fig. 9. For a given distribution of conductivity with depth, $\sigma(z)$, the corresponding induction log $V(z)$ obtained with the logging probe being modelled is given by

$$V(z) = \sigma(z) * g(z) \tag{21}$$

To the extent that the assumption of a linear system is valid, it is clear that the system response function may be removed to leave the conductivity–depth profile if a suitable stable inverse filter can be derived. In addition, the response of a given logging probe can be modified to simulate the response of a different probe in a manner analogous to that proposed for acoustic logs by Foster et al.,[14] discussed in Section 3. George et al.[17] used frequency-domain division to derive digital filters for these purposes.

In the case of resistivity logs, altering the apparent detector spacing by digital filtering obviously does not change the actual depth of investigation

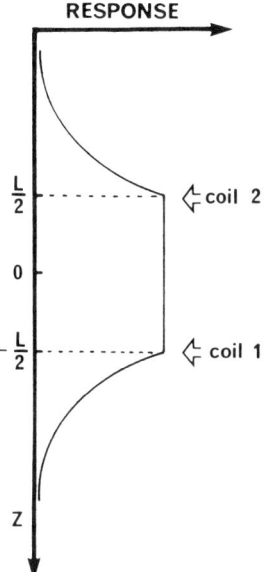

FIG. 9. Idealised geometric factor (spatial impulse response) for a 2-coil induction probe.[18a]

of the probe. Thus, to the extent that these 'response equalisation' techniques produce the intended results, the log made with a given spacing and the synthetic log processed to give the same apparent spacing will agree only in the absence of radial conductivity variations. George et al.[17] suggested that this fact can be used to give improved information on the invaded zone. By processing both a high resolution (short-spacing) resistivity log and a medium-spacing induction log to simulate the characteristics of a long-spacing induction log, the 3 logs can be employed as usual to give information on the invaded zone. Using the processed logs rather than the raw logs offers the advantage (ideally) of reducing or eliminating one variable—vertical spatial resolution—from the logs being compared.

George et al.[17] presented computer simulations using artificially generated log data to illustrate the above techniques, along with an iterative method for computing the depth of invasion. These data were generated under the same simplifying assumptions as the digital filter techniques used by those authors. They did not show examples of actual field logs to demonstrate their techniques.[17]

Howell and Fisher[18b] used a similar conceptual approach to deconvolving induction logs. They used least square error inverse filters, with a geometric factor model for the input function, and various desired output functions depending on the signal-to-noise characteristics of the data. The main thrust of their work[18b] was concerned with deconvolving induction logs from deviated boreholes. They used a simple model to derive filters designed to convert a log from a deviated borehole to the equivalent log from a vertical hole, as well as straightforward inverse filters to convert a log from a deviated or vertical borehole into the corresponding resistivity–depth profile. The authors presented several examples with simulated and actual data to illustrate the techniques.

The application of linear digital filters to induction logs involves the implicit assumption that the problem is a linear one. In ignoring propagation and skin effects, George[16] remarked that for a resistivity of 5 ohm-m and a 1 m coil spacing, an error of not more than 10% is introduced. In some geological situations, sequences having resistivities an order of magnitude lower than this are common. In these cases, the assumption of linearity is not a good one, although the techniques may still be of some use qualitatively. A discussion of the problem of skin effect in induction logging was presented by Gianzero and Anderson.[19] Figure 10,

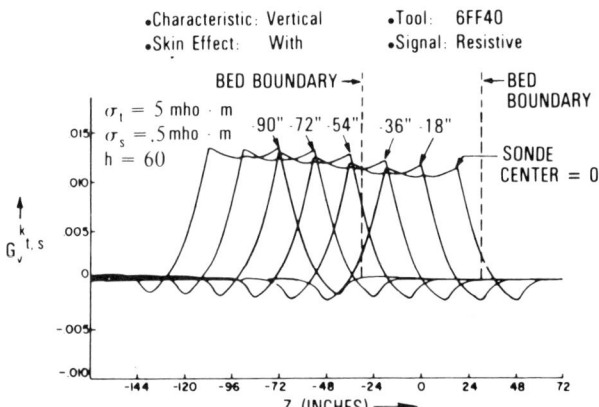

FIG. 10. Variation of system response function for a focussed induction probe in the vicinity of a bed of contrasting conductivity.[19] The ordinate represents the vertical resistive tool response ($G_v^{k,t,s}$), and the abscissa represents depth. An infinite thickness of material having conductivity $\sigma_s = 0.5$ mho/m is interrupted by a bed having thickness $h = 60$ in and conductivity $\sigma_t = 5$ mho/m. For more details see Gianzero and Anderson.[19]

reproduced from that paper, shows the resistive vertical characteristics ($G_v^{k_{1,s}}$) of a particular focussed induction device as the probe moves past a 60 in (1·5 m) bed having conductivity 5 mho/m, embedded in an infinite medium having conductivity 0·5 mho/m. Careful examination of Fig. 10 shows a change in the probe-response characteristics with position relative to the bed; a linear device would have constant response characteristics.

5. TEMPERATURE LOGS

Figure 11 shows a temperature log of a thermally stable borehole. Aside from the temperature inversion in the top 80 m or so, due to surface temperature effects, remarkably little character is apparent in the log. Looking at Fig. 11 more closely, especially by sighting along the curve, it is possible to discern a number of small irregularities in what at first glance appears to be a nearly straight line. In the case of uniform heat flow along a

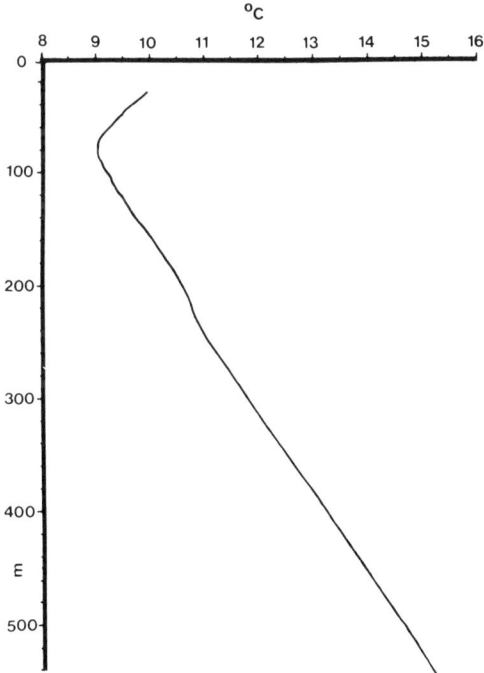

FIG. 11. Temperature log from a thermally stable borehole made with a high precision (± 0.001 °C) digital temperature logging system.

thermally stable borehole these variations in temperature *gradient* (first derivative of temperature with respect to depth) are in theory proportional to the thermal resistivity of the rock strata (see Diment,[20] Gretener[21] and Sammel[22] for a discussion of thermal stability). In other words, there is a great deal of fine detail in a temperature log which under favourable conditions may be useful, but which is very difficult to discern. Over the years a number of proposed solutions to the problem of utilising this hidden information have been tried. The logs produced have often been crude, noisy or highly qualitative in nature.

Costain[23] presented the results of computer model studies in which he applied digital-filtering techniques to computer-simulated temperature logs. Costain showed that simple digital filters could be used to remove the smearing effect of the thermistor's exponential time constant, and numerically differentiate the temperature data to give a deconvolved temperature gradient log. Costain used the Fourier transform to derive the inverse filter for the thermistor time constant, as described here. The thermistor impulse response $r(t)$ is approximately exponential of the form

$$r(t) = 0 \qquad \text{for } t < 0$$

$$r(t) = \frac{1}{T}\exp(-t/T) \qquad \text{for } t \geq 0 \qquad (22)$$

where t is time and T is the thermistor 1/exp time constant (the time required for a thermistor subjected to a step temperature change to reach $(1 - 1/\exp)$ of the final temperature). The Fourier transform $R(\omega)$ of $r(t)$ is given by:

$$R(\omega) = \int_{-\infty}^{\infty} r(t)\exp(-j\omega t)\,dt = \frac{1}{1+j\omega T} \qquad (23)$$

This expression is then inverted to give the Fourier transform $F(\omega)$ of the desired time-domain inverse filter $f(t)$, or

$$F(\omega) = 1 + j\omega T \qquad (24)$$

and

$$f(t) = \frac{1}{2\pi}\int_{-\infty}^{\infty} (1+j\omega T)\exp(j\omega t)\,d\omega = \delta(t) + T\delta'(t) \qquad (25)$$

where $\delta(t)$ is the Dirac delta function, $\delta'(t)$ is its first derivative, and $j = \sqrt{-1}$. Costain then approximated $\delta'(t)$ using a geometrical argument to give the approximate digital inverse filter. Working from the same exact analytical solution for continuous data (eqn (25)) Conaway[24] followed a somewhat different path to derive the approximate inverse filter. According to this latter approach an actual temperature function $\theta(t)$ is related to the measured output of the temperature sensor $\theta_m(t)$ by

$$\theta(t) = f(t) * \theta_m(t) = \int_{-\infty}^{\infty} (\delta(\tau) + T\delta'(\tau))\theta_m(t-\tau) \, d\tau \qquad (26)$$

Skipping the intermediate steps[24] leads to

$$\theta(t) = \theta_m(t) + T\theta'_m(t) \qquad (27)$$

Approximating $\theta'_m(t)$ (the first derivative of the measured temperature function) for digital data for the nth reading by the first symmetrical difference

$$\frac{\theta_m|_{n+1} - \theta_m|_{n-1}}{2\Delta t} \qquad (28)$$

and applying this approximation to eqn (27) leads to the 3-coefficient approximate digital inverse filter

$$\left[\frac{T}{2\Delta t}, 1, -\frac{T}{2\Delta t} \right] \qquad (29)$$

where Δt is the sampling interval. For constant logging velocity v the conversion of eqn (29) to an inverse filter $f(z)$ based on depth z is, of course, trivial. Since $\Delta t = \Delta z/v$, the inverse filter coefficients become

$$\left[\frac{vT}{2\Delta z}, 1, -\frac{vT}{2\Delta z} \right] \qquad (30)$$

where Δz is the depth sampling interval.

Equations (29) and (30) give the desired inverse filter in terms of time and depth, respectively. A second approximate filter based on eqn (28) was suggested by Costain[23] to give a temperature gradient profile by numerical differentiation of the temperature log using the first symmetrical difference. This filter $g(z)$ is given by the three coefficients:

$$\left[\frac{1}{2\Delta z}, 0, -\frac{1}{2\Delta z} \right] \qquad (31)$$

In practice, high frequency noise is greatly amplified by application of the above deconvolution and differentiation filters, and some sort of

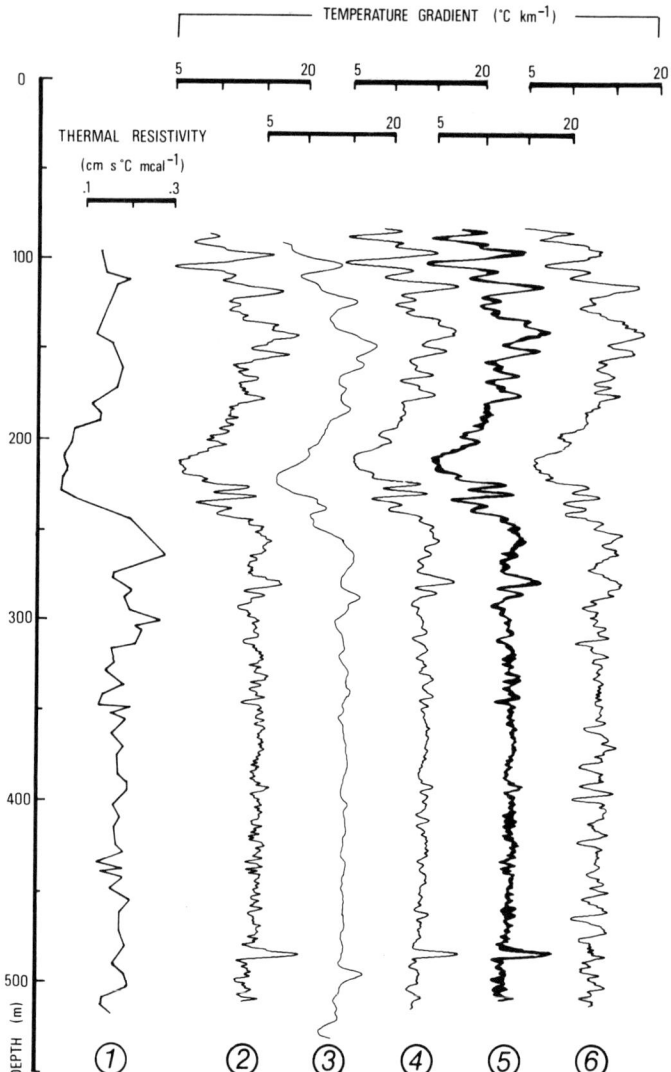

FIG. 12. Curve 1, thermal resistivity profile of the borehole corresponding to Fig. 11, based on laboratory measurements on core samples; curve 2, deconvolved temperature gradient log made at a logging speed of 18 m/min down the hole; curve 3, same data as in curve 2, but produced without deconvolution; curve 4, deconvolved temperature gradient log made at 9 m/min down hole; curve 5, profile made by superimposing curves 2 and 4, and filling in the spaces between the lines, to show repeatability; curve 6, deconvolved temperature gradient log made at 18 m/min up hole.[24]

smoothing technique must be used. An approach suggested by Costain [23] and further investigated by Conaway [24] involves stretching the deconvolution and differentiation filters to include more temperature readings. This gives the resulting digital filters less gain at high frequencies. A somewhat simpler approach is to use the 3-coefficient digital filters for deconvolution and differentiation (eqns (30) and (31)) and a simple low-pass digital filter as described earlier in Section 2.4.

The temperature log data shown in Fig. 11 were processed [24] to give a deconvolved temperature gradient profile of the borehole, as shown in Fig. 12, curve 2. Comparison with curve 1, the thermal resistivity profile of the borehole from laboratory measurements on core samples, shows a similarity in overall form, although curve 2 has more fine detail. Under the conditions described earlier, the temperature gradient profile is essentially proportional to the thermal resistivity profile. Curve 3 shows the same data differentiated and smoothed as in curve 2, but without applying the inverse filter (eqn (30)). Much of the fine detail is missing or degraded in curve 3 and the log features are displaced in the direction of probe motion (in this case, down the hole). Curve 4 shows another log of the borehole at a slower speed, resulting in somewhat lower noise. Curve 5 shows curves 2 and 4 superimposed with the spaces between the curves filled in to give an indication of repeatability, which is quite good. Curve 6 shows a processed log similar to curves 2 and 4, except that it was made while withdrawing the probe from the hole. There is quite a bit more noise in this log due to temperature disturbances caused by the logging probe and cable. It is clear that for best results the hole should be logged on a downward pass, although this is not always practical. This also explains why temperature gradient logs produced by taking the temperature differential between two sensors, one at the top and one at the bottom of the probe, are inherently noisy and inaccurate.

An application of temperature gradient logs obtained by applying digital filters to high precision temperature logs as described above was given by Conaway and Beck.[25] Under favourable conditions thermal gradient logs can give lithologic information (based on thermal resistivities) with the advantage that they can be run in cased holes, unlike electrical resistivity logs. An example of this lithologic correlation is shown in Fig. 13; thermal resistivity values based on core samples from this borehole (Fig. 13) are given in Table 1.

The deconvolution and differentiation of temperature logs is a good example of applying digital filtering techniques to enhance or amplify useful information which is present, but hidden, in the log.

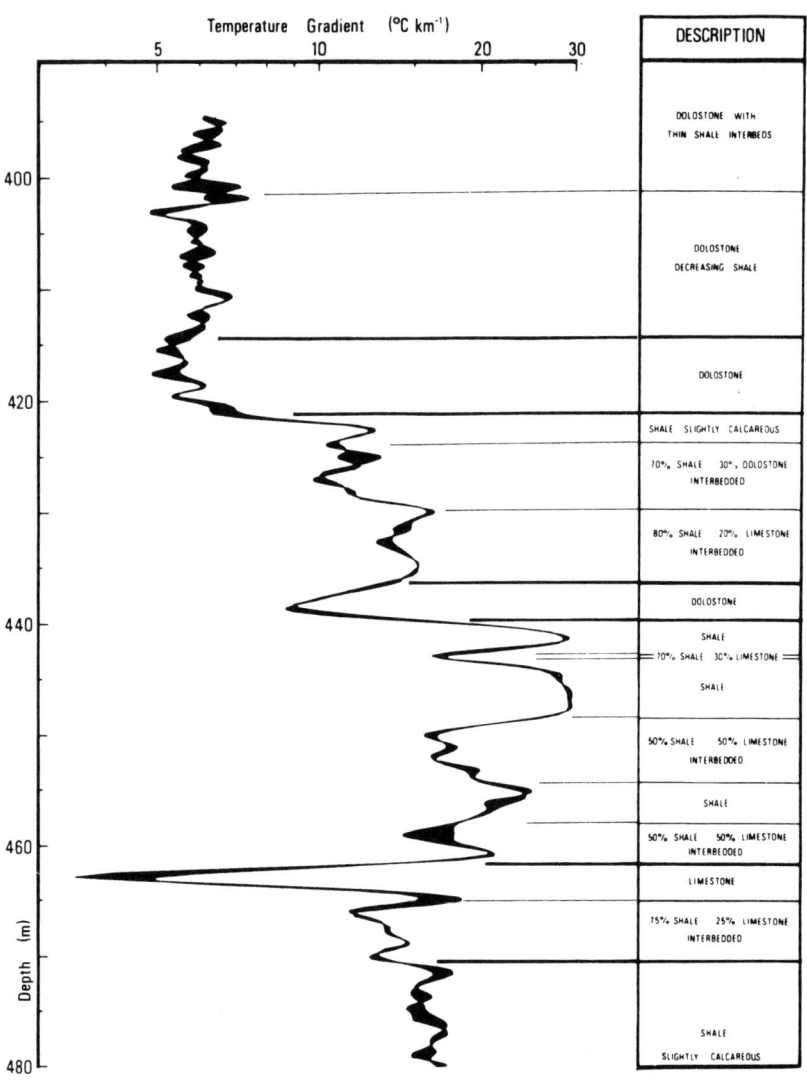

FIG. 13. Correlation of a high precision temperature gradient log with a geologic log of the core.[25]

TABLE 1
VALUES OF THERMAL RESISTIVITIES BY ROCK TYPE AND FORMATION FOR THE BOREHOLE CORRESPONDING TO THE LOGS SHOWN IN FIG. 13.[25]

Rock Type	Geologic Symbol Used	Formation	Thermal Resistivity Mean	Standard Deviation	Number of Samples
LIMESTONE		DETROIT RIVER	0.143	0.019	21
		BOIS BLANC	0.119	0.015	7
		ROCHESTER	0.150	....	2
		CABOT HEAD	0.129	....	1
		MANITOULIN	0.163	0.032	3
		ALL FORMATIONS	0.140	0.022	34
DOLOSTONE		DETROIT RIVER	0.091	....	1
		BASS ISLANDS	0.098	0.020	9
		SALINA	0.088	0.006	9
		REYNALES	0.116	....	2
		GUELPH	0.086	0.005	7
		LOCKPORT	0.087	0.004	13
		ALL FORMATIONS	0.092	0.013	34
SHALE		SALINA	0.168	0.041	6
		ROCHESTER	0.202	0.014	4
		CABOT HEAD	0.260	0.062	6
		QUEENSTON	0.202	0.018	21
		ALL FORMATIONS	0.206	0.041	37
ANHYDRITE		SALINA	0.072	0.002	4
GYPSUM		ALL FORMATIONS	0.250	0.033	3

6. ANALOG RATEMETER

Analog logging systems often employ a ratemeter in the output to the chart recorder. This is an electronic circuit which smooths the output and, in the case of nuclear logs, integrates individual pulses to give the count rate or radiation intensity. If the analog chart recording is to be digitised and processed it may be desirable to remove the smearing effect of the ratemeter at the same time (see Fig. 14). This technique rarely approaches the quality normally obtained by a true digital logging system, but may offer some improvement.

Although various analog filters are used for smoothing the log in real time, probably the most common type has an exponential impulse response $r(t)$ which is the same as the response given by eqn (22) for a temperature sensor. The same 3-coefficient approximate inverse filter developed earlier (eqn (30)) obviously can be used here as well. However, it has been shown

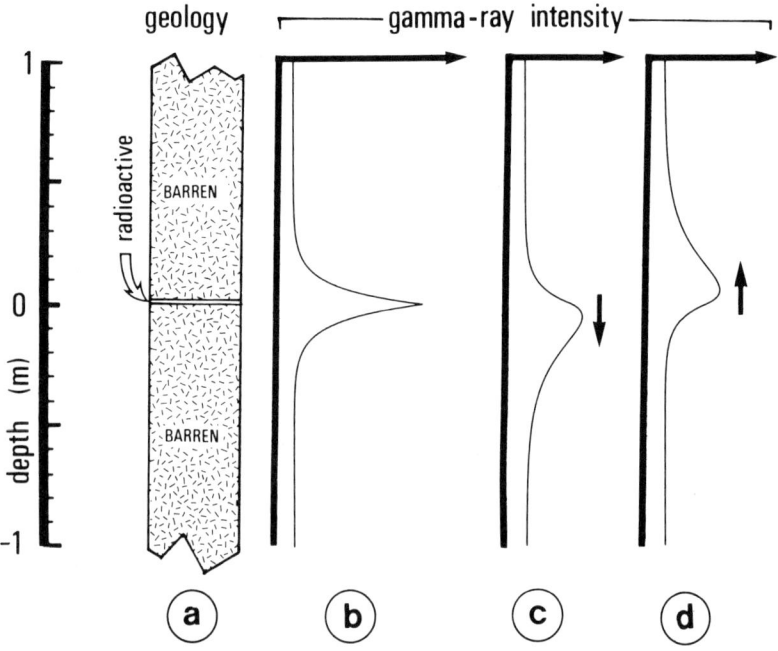

FIG. 14. (a) Geologic column showing a thin radioactive zone embedded in barren rock. (b) Ideal gamma-ray log past the thin radioactive zone, with no ratemeter effect. (c) Ideal response of an analog logging system downward past the zone. (d) As in (c) except logging upward.[26]

by Conaway[26] that an exact inverse filter derived using **Z**-transform techniques can give improved results over the approximate filter. The derivation begins by expressing the impulse response of the ratemeter $r(t)$ in discrete form:

$$\ldots, 0, 0, \frac{1}{vT}, \frac{1}{vT}\exp(-\Delta z/vT), \frac{1}{vT}\exp(-2\Delta z/vT), \ldots \quad (32)$$

or

$$R(\mathbf{Z}) = \frac{1}{vT} + \frac{\mathbf{Z}^{-1}}{vT}\exp(-\Delta z/vT) + \frac{\mathbf{Z}^{-2}}{vT}\exp(-2\Delta z/vT) + \cdots \quad (33)$$

where $R(\mathbf{Z})$ is the **Z**-transform of the digital ratemeter impulse response. Equation (33) may be rewritten as

$$R(\mathbf{Z}) = \sum_{n=0}^{\infty} \frac{\mathbf{Z}^{-n}}{vT} \exp(-n\Delta z/vT) \quad (34)$$

We wish to find the inverse filter $F(\mathbf{Z})$ such that

$$F(\mathbf{Z}) \cdot R(\mathbf{Z}) = 1 \quad (35)$$

Therefore

$$F(\mathbf{Z}) = \frac{1}{R(\mathbf{Z})} \quad (36)$$

substituting eqn (34) into eqn (36) and dividing gives

$$F(\mathbf{Z}) = vT - \mathbf{Z}^{-1} vT \exp(-\Delta z/vT) \quad (37)$$

This is a 2-coefficient exact inverse filter

$$[vT, -vT\exp(-\Delta z/vT)] \quad (38)$$

or in the normalised form

$$\left[\frac{1}{(1-\exp(-\Delta z/vT))}, -\exp\frac{(-\Delta z/vT)}{(1-\exp(-\Delta z/vT))}\right] \quad (39)$$

As an example of the improvement which can be gained by using this exact inverse filter rather than the approximate filter (eqn (30)) consider Fig. 15. Two gamma-ray logs run with an analog system past the same radioactive zone at the same logging speed, but in different directions are shown together as Fig. 15(a). The smearing of the features in the direction of probe motion is obvious here. In Fig. 15(b), the approximate inverse

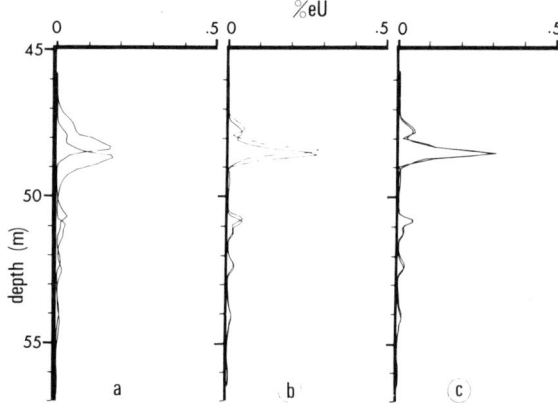

FIG. 15. (a) Two gamma-ray logs run in opposite directions past the same geologic sequence, with $v = 6$ m/min, $T = 2 \cdot 6$ s, and $\Delta z = 10$ cm. (b) Logs shown in (a) after application of the approximate inverse filters, as described in Section 6. (c) Logs shown in (a) after application of the exact inverse filters.[26]

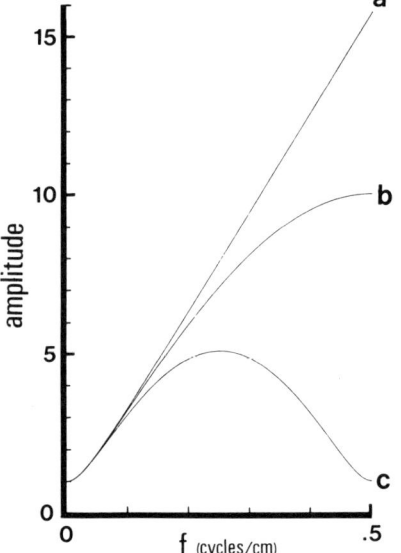

FIG. 16. (a) Amplitude spectrum of the theoretical continuous inverse filter for $vT = 5$ cm. (b) Amplitude spectrum of the corresponding exact digital inverse filter for $\Delta z = 1$ cm. (c) Amplitude spectrum of the approximate digital filter for $\Delta z = 1$ cm.[26]

filter for the exponential ratemeter (eqn (30)) has been applied, as has the inverse filter for gamma-ray logs (Section 7). Spatial resolution has been improved, and the depth offset has been reduced. In Fig. 15(c) the exact ratemeter inverse filter (eqn (39)) has been applied rather than the approximate filter. The agreement between the two logs is greatly improved over the unprocessed logs (Fig. 15(a)) and the depth registration is now very good.

It is instructive to look at these inverse ratemeter filters in the frequency domain. The amplitude spectrum of the theoretical analog inverse filter (eqn (25)) is shown as curve a in Fig. 16 for $vT = 5$ cm. The amplitude spectrum of the corresponding exact digital inverse filter (eqn (39)) is shown as curve b in this figure, while the amplitude spectrum of the approximate inverse filter (eqn (30)) is given by curve c. The difference between the exact and approximate filters (curves b and c) results from approximating the first derivative by the first symmetrical difference in the derivation of the approximate filter. For this same reason the performance of the approximate filter improves as the sampling interval decreases. In fact, the three filters (analytical, exact digital and approximate digital) as well as the LSE filter will converge as Δz approaches zero. The amplitude spectrum of the exact digital filter (curve b) does not match that of the theoretical analog filter (curve a) because of aliasing in the digitising process. Because the ratemeter impulse response is not band-limited (i.e. has no upper bound on frequency content) aliasing is unavoidable unless the signal is filtered with a low-pass analog filter before digitising.

7. GAMMA-RAY LOGS

The application of digital filtering techniques to gamma-ray logs has been reviewed in an earlier volume of this series by Killeen,[27] and elsewhere by Conaway.[28] Because of the wide application of gamma-ray logs in mining and petroleum exploration, the major points will be reiterated here for completeness.

A thin zone of radioactive material (Fig. 14(a)) gives rise to a log anomaly which is spread along the borehole (Fig. 14(b)). Figure 14(b) represents the ideal noise-free system response function; the desired output of the deconvolution operation is a rectangle equal in width to the thickness of the radioactive zone, with amplitude proportional to the concentration or grade G of radioactive material in the zone. Scott et al.[29] showed that given certain assumptions (such as constant gamma-ray attenuation characteristics with depth), linear processing techniques may be applied to this problem.

One approach to deconvolving the gamma-ray log is to use an optimum inverse filter. The impulse response of the system may be determined by logging past a radioactive zone (ideally having thickness equal to the sampling interval) either in a model borehole or a field borehole that has been cored and analysed. Alternatively, a log can be run past a step change in radioactivity and the log differentiated numerically[12,30] to give the impulse response (Fig. 17(a)). The desired output is then specified to be an impulse (Fig. 17(b)) and an optimum inverse filter (in this case a least square error filter) of a specified length is generated (Fig. 17(c)) as described in Section 2.3. When applied to the system impulse response (Fig. 17(a)), the actual output of the filter is an impulse with some noise (Fig. 17(d)). If the desired output is specified as a smoothing filter of some sort then the

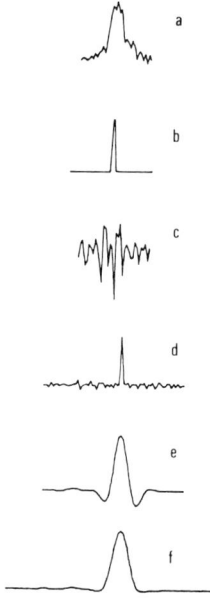

FIG. 17. Illustration of the steps in computing LSE inverse filters: (a) experimentally determined system impulse response; (b) desired output, an impulse; (c) LSE filter; (d) actual output obtained by convolving (a) with (c); (e) LSE filter to convert (a) into a smoothed output function; (f) actual output obtained by convolving (a) with (e).

resulting optimum inverse filter in this case looks like Fig. 17(e) and produces the actual output shown in Fig. 17(f) when convolved with the input function (Fig. 17(a)). The actual filter which would be chosen would probably fall between Fig. 17(c) (too noisy) and Fig. 17(e) (excessive smoothing).

It has been shown that the shape of the system response function depends on many factors, including (among others) formation bulk density, borehole diameter, borehole fluid density, casing, and instrumental parameters including detector characteristics, ratemeter time constant (if any) and sampling interval.[31-37] Thus, a given optimum inverse filter will only be completely valid as long as those parameters remain unchanged. This means that a new input function should be determined, and a new filter generated for each new combination of borehole, formation, and instrumental parameters encountered. As an example, Fig. 18 shows the variation in the shape of the ideal system response function across a boundary which represents a density contrast in the rock.

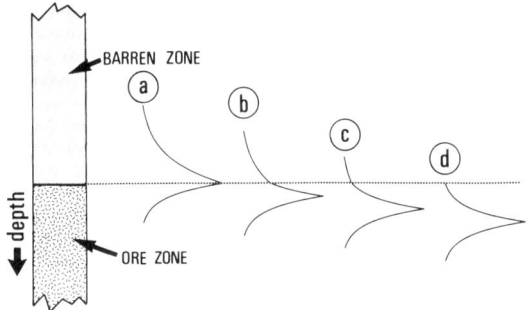

FIG. 18. Change in the shape of the ideal gamma-ray logging system response function from a low density zone (top) to a high density zone (bottom). Curves (a)–(d) represent the system impulse response at a different position with respect to the barren zone–ore zone interface, as shown.[36]

A modification of the above procedure involves determining the system response function analytically rather than experimentally. If this could be accomplished accurately, then an optimum inverse filter could be generated fairly easily for each new set of conditions encountered. Analytical solutions have been proposed by Suppe[38] and by Czubek.[32,33,39] For example, Czubek's analytical expression for the response ϕ_0 of a point detector to a thin zone of radioactive material at depth $z = 0$ may be written as:[39]

$$\phi_0(\mu, z) = \frac{E_1[\mu\sqrt{R^2 + z^2}]}{2\mu R[K_1(\mu R) - \int_{\mu R}^{\infty} K_0(x)\,dx]} \qquad (40)$$

where μ is the linear attenuation coefficient, R is the borehole radius, $K_0(x)$ and $K_1(x)$ are modified Bessel functions of the second kind, and $E_1(x)$ is the exponential integral of order 1 defined by

$$E_1(x) = \int_x^{\infty} \frac{\exp(-t)}{t}\,dt \qquad (41)$$

Equation (40) neglects scattering of the radiation, borehole fluid, casing, eccentric position of the probe in the hole, and the spatial and energy response characteristics of the logging system. As these factors are folded into the analytical system response function (eqn (40)), the complexity of the expression increases greatly as does the number of physical parameters which must be determined before the analytical expression can be evaluated numerically. It is clear that this approach is impractical for general use.

Davydov[40] developed a deconvolution scheme for gamma-ray logs in which the system response function is approximated by the double-sided exponential function $\phi(z)$ where

$$\phi(z) = \frac{\alpha}{2}\exp(-\alpha|z|) \qquad (42)$$

where the shape constant α (not to be confused with the spectral stripping factor α) is a function of many borehole, formation, and instrumental parameters. It has been shown[35,36] that the best results using this approximation are obtained if the value of α is determined by plotting the natural logarithm of the amplitude of the system impulse response as a function of depth; the slope of the linear (or nearly linear) flanks of the anomaly gives α directly (Fig. 19). This method may also be used under favourable conditions in field boreholes as described by Conaway[35] (Fig. 20).

An approximate inverse filter may be developed based on eqn (42) using the Fourier transform approach, much as was shown earlier for the thermistor response in temperature logs.[12,40] This filter has 3 coefficients given by

$$\left[-\frac{1}{(\alpha\Delta z)^2},\ 1 + \frac{2}{(\alpha\Delta z)^2},\ -\frac{1}{(\alpha\Delta z)^2}\right] \qquad (43)$$

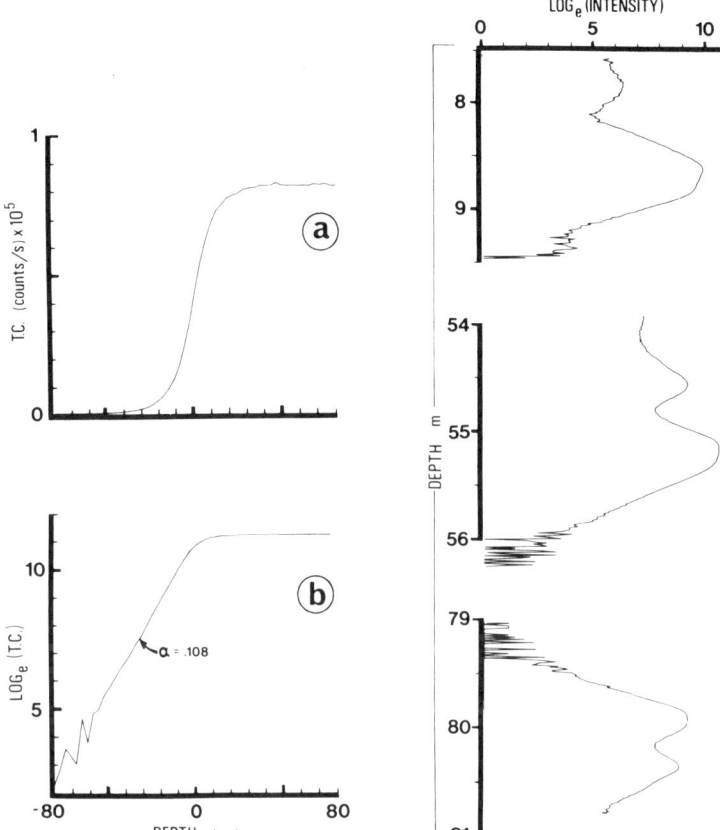

FIG. 19. (a) Log from a barren zone into a radioactive zone in a model borehole. (b) The semi-logarithmic plot of (a), indicating that α (given here in units of cm^{-1}) is given by the slope of the anomaly flank outside the radioactive zone.[36]

FIG. 20. Three segments of a gamma-ray log, plotted background subtracted on a semi-logarithmic scale. The value of α may be determined from such field logs where there is a distinct transition from a barren zone to a radioactive zone, as shown in Fig. 19.[35]

This inverse filter will significantly improve the spatial resolution of a gamma-ray log in many cases as shown in Fig. 21. Some residual distortion will remain in the processed log due to the fact that eqn (42) is only an approximation of the actual system response function. This is illustrated in Fig. 22. In particular, the longer the detector or the larger the borehole, the more distortion will remain. In practice it is difficult to remove this residual distortion because of limitations imposed by noise.

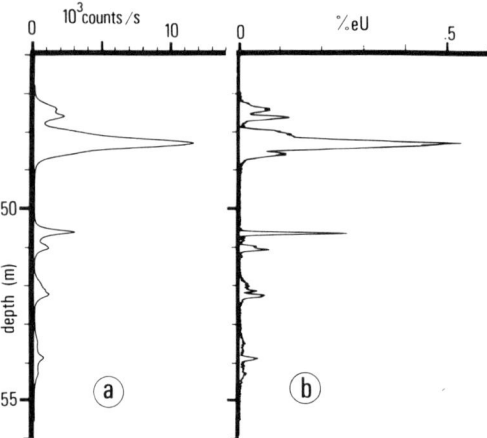

FIG. 21. Digital gamma-ray log using a 25×25 mm detector, $\Delta z = 1$ cm, $v = 0.3$ m/min. Three runs have been plotted superimposed to show repeatability. (b) Same 3 logs as in (a), deconvolved. From Conaway et al.[42]

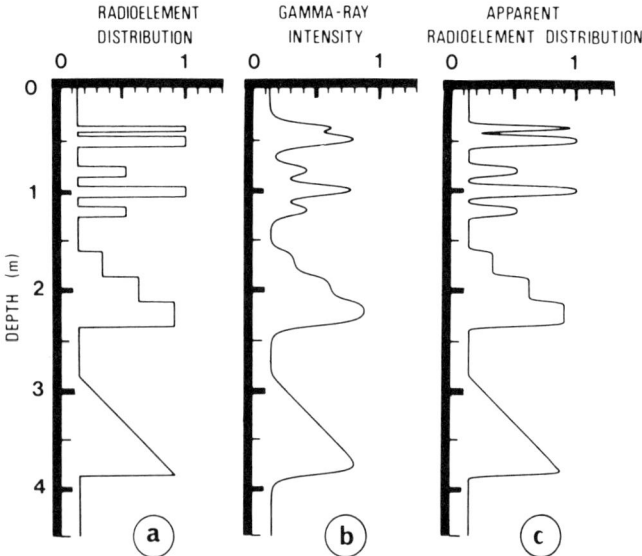

FIG. 22. (a) Computer-generated radioelement distribution. (b) Corresponding noise-free gamma-ray log obtained by convolving (a) with an analytical response function based on eqn (40). (c) Apparent radioelement distribution obtained by deconvolving (b) with the 3-coefficient inverse filter (eqn (43)).[36]

The approximate method using the filter given as eqn (43) is relatively simple to use, and yet is flexible. With some practice the log analyst can often tell if an incorrect value of α has been used in the processing; if α is too large, little change will be noted in the processed log, while too small a value of α will cause overshoot and negative radioelement concentrations. In the case of a long detector or large diameter borehole, an optimum filter may prove to be more effective, albeit more difficult to use.

8. GENERALISED RESPONSE EQUALISATION FILTERING

Branisa[41] applied the idea of response equalisation filtering to well logs in general. He presented a qualitative approach wherein any two or more logs could be filtered to have similar frequency characteristics; thereby, he argued, making them easier to compare and correlate. This was a generalisation of the approach suggested by Foster et al.[14] for changing the effective spacing of acoustic logs, and by George et al.[17] for equalising the effective vertical resolution of induction and electric resistivity logs of various spacings.

As an example of the application of this technique, Fig. 23,[41] shows an SP log and a gamma-ray log from two wells in the same area. The procedure is

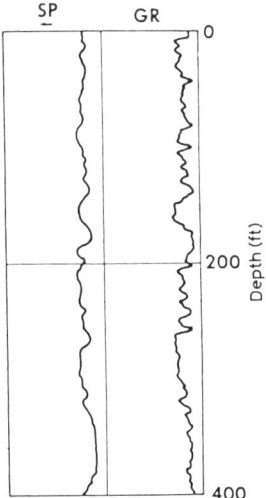

FIG. 23. SP and gamma-ray logs from two wells in the same area (vertical axis is depth in feet).[41]

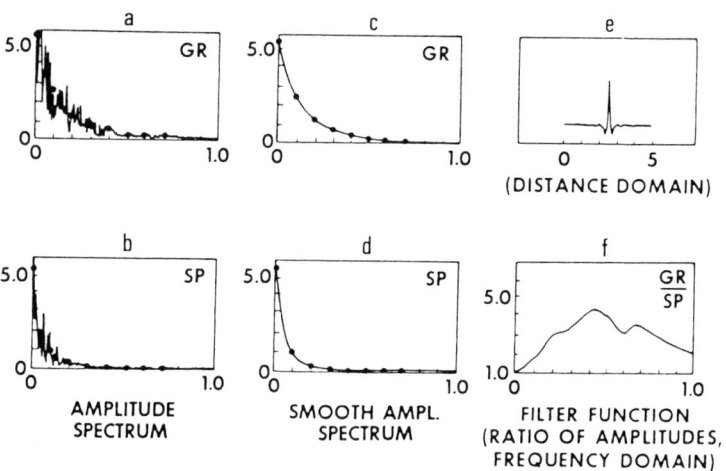

FIG. 24. Procedure for computing the response equalisation filter:[41] (a) and (b) represent the amplitude spectra for a section of a gamma-ray log and of the corresponding SP log, respectively; (c) and (d) are the corresponding spectra 'smoothed' or simplified by eye; (e) is the impulse response of the response equalisation filter and (f) is its amplitude spectrum.

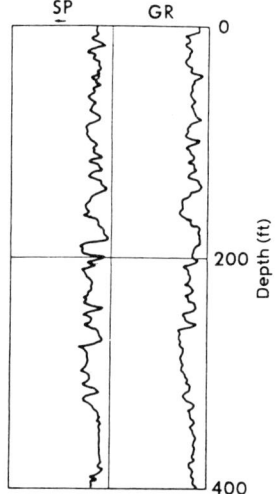

FIG. 25. Same two logs as in Fig. 23, with the exception that the SP curve has had the response equalisation filter applied.[41]

first to compute the frequency spectrum of a sample log with each probe past the same geologic sequence, using a discrete Fourier transform (Figs 24(a) and (b)). Each spectrum is then simplified or 'smoothed' by eye (Figs 24(c) and (d)). The resulting smoothed spectra can be divided one into the other (Fig. 24(f)) and the result transformed to give the desired 'distance domain' response equalisation filter (Fig. 24(e)). Application of this digital filter to the SP log shown in Fig. 23 gives the filtered SP curve in Fig. 25 plotted with the same unprocessed gamma-ray curve. The two curves now exhibit similar frequency characteristics, which may facilitate interpretation or correlation in some situations.

9. CONCLUSIONS

Linear digital filters can be used to change the frequency characteristics of geophysical logs in many ways. Regardless of whether a log is being processed to extract accurate quantitative information, or to change its appearance qualitatively to facilitate interpretation, it is important that the assumptions inherent in developing the filters be carefully examined. For example, developing an inverse filter for induction logs based on the geometric factor approach neglects propagation and skin effects. For formation resistivities less than several ohms this may result in significant error. Similarly, filtering of gamma-ray logs to determine the distribution of radioactive material with depth along the borehole is only valid if the gamma-ray attenuation characteristics of the formations through which the borehole passes are constant. Filtering temperature logs to give a formation thermal resistivity profile requires a condition of uniform heat flow along the borehole.

Rarely will these conditions, and other similar implicit assumptions, be satisfied completely in nature. The log analyst must understand the limitations that these assumptions impose on the accuracy and reliability of the processed log, and use the available processing techniques judiciously to improve or facilitate the interpretation.

REFERENCES

1. LINDSETH, R. O. (1966) Application of signal theory to well log interpretation. In *SPWLA Transactions of the 7th Annual Logging Symposium*, Society of Professional Well Log Analysts, Houston, Texas.
2. BRACEWELL, R. N. (1978) *The Fourier Transform and Its Applications*, McGraw–Hill, New York.

3. KANASEWICH, E. R. (1973) *Time Sequence Analysis in Geophysics*, University of Alberta Press, Calgary, Alberta.
4. BRIGHAM, E. O. (1974) *The Fast Fourier Transform*, Prentice–Hall, Englewood Cliffs, New Jersey.
5. ROBINSON, E. A. (1967) *Multichannel Time Series Analysis with Digital Computer Programs*, Holden–Day, San Francisco, California.
6. IEEE (1979) *Programs for Digital Signal Processing*, IEEE Press, John Wiley and Sons, New York.
7. TREITEL, S. and ROBINSON, E. A. (1964) The stability of digital filters. *IEEE Trans. Geosci. Electr.* **GE-2**, 6–18.
8. JURY, E. I. (1964) *Theory and Application of the Z-transform Method*, John Wiley and Sons, New York.
9. SHANKS, J. L. (1967) Recursion filters for digital processing. *Geophysics* **32**, 33–51.
10. RICE, R. B. (1962) Inverse convolution filters. *Geophysics* **27**, 4–18.
11. ROBINSON, E. A. and TREITEL, S. (1967) Principles of digital Wiener filtering. *Geophys. Prosp.* **15**, 311–33.
12. CONAWAY, J. G. and KILLEEN, P. G. (1978) Quantitative uranium determinations from gamma ray logs by application of digital time series analysis. *Geophysics* **43**, 1204–21.
13. SHULTZ, W. E. and THADANI, S. G. (1981) Applications of digital filtering techniques to nuclear well logs. In *SPWLA Transactions of the 22nd Annual Logging Symposium*, Society of Professional Well Log Analysts, Houston, Texas.
14. FOSTER, M. R., HICKS, W. G. and NIPPER, J. T. (1962) Optimum inverse filters which shorten the spacing of velocity logs. *Geophysics* **27**, 317–26.
15. RUNGE, R. J. and POWELL, N. J. (1967) The effect of sampling rate on sonic log span adjustment. In *SPWLA Transactions of the 8th Annual Logging Symposium*, Society of Professional Well Log Analysts, Houston, Texas.
16. GEORGE, JR., C. F. (1963) Application of special filtering techniques to well log analysis. Doctoral dissertation, The University of Texas, Austin, Texas.
17. GEORGE, JR., C. F., SMITH, H. W. and BOSTICK, F. X. (1964) Application of inverse filters to induction log analysis. *Geophysics* **29**, 93–104.
18(a). DOLL, H. G. (1949) Introduction to induction logging and application to logging of wells drilled with oil base mud. *J. Pet. Tech.* **1**, 148–62.
18(b). HOWELL, E. P. and FISHER, T. E. (1982) Induction log deconvolution for deviated boreholes. In *SPWLA Transactions of the 23rd Annual Logging Symposium*, Society of Professional Well Log Analysts, Houston, Texas.
19. GIANZERO, S. and ANDERSON, B. (1982) A new look at skin effect. *The Log Analyst* **23**(1), 20–34.
20. DIMENT, W. H. (1967) Thermal regime of a large diameter borehole: instability of the water column and comparison of air- and water-filled conditions. *Geophysics* **32**, 720–6.
21. GRETENER, P. E. (1967) On the thermal instability of large diameter wells—an observational report. *Geophysics* **32**, 727–38.
22. SAMMEL, E. A. (1968) Convective flow and its effect on temperature logging in small-diameter wells. *Geophysics* **33**, 1004–12.

23. Costain, J. K. (1970) Probe response and continuous temperature measurements. *J. Geophys. Res.* **75**, 3968–75.
24. Conaway, J. G. (1977) Deconvolution of temperature gradient logs. *Geophysics* **42**, 823–37.
25. Conaway, J. G. and Beck, A. E. (1977) Fine scale correlation between temperature gradient logs and lithology. *Geophysics* **42**, 1401–10.
26. Conaway, J. G. (1980) Exact inverse filters for the deconvolution of gamma ray logs. *Geoexploration* **18**, 1–14.
27. Killeen, P. G. (1982) *Developments in Geophysical Exploration Methods—3*, Ed. A. A. Fitch, Applied Science Publishers, London, pp. 95–150.
28. Conaway, J. G. (1982) Principles of inverse filtering applied to gamma ray logs. Proceedings of the EAEA Conference on Uranium Exploration, Paris, 1982.
29. Scott, J. H., Dodd, P. H., Droullard, R. F. and Mudra, P. J. (1961) Quantitative interpretation of gamma-ray logs. *Geophysics* **26**, 182–91.
30. Scott, J. H. (1963) Computer analysis of gamma-ray logs. *Geophysics* **28**, 457–65.
31. Rhodes, D. F. and Mott, W. E. (1966) Quantitative interpretation of gamma-ray spectral logs. *Geophysics* **28**, 410–18.
32. Czubek, J. A. (1961) Some problems of the theory and quantitative interpretation of the gamma-ray logs. *Acta Geophysica Polonica* **9**, 121–37.
33. Czubek, J. A. (1962) The influence of the drilling fluid on the gamma-ray intensity in the borehole. *Acta Geophysica Polonica* **10**, 25–30.
34. Czubek, J. A. (1969) Influence of borehole construction on the results of spectral gamma-logging. In *Nuclear Techniques and Mineral Resources*, IAEA Proceedings Series, IAEA, Vienna.
35. Conaway, J. G. (1980) Direct determination of the gamma ray logging system response function in field boreholes. *Geoexploration* **18**, 187–99.
36. Conaway, J. G. (1980) Uranium concentrations and the system response function in gamma ray logging. In *Current Research*, Geol. Sur. Can., Paper 80-A, pp. 77–87.
37. Wilson, R. D. and Stromswold, D. C. (1981) Spectral gamma ray logging studies. Report Bendix Field engineering Corp. for USDOE.
38. Suppe, S. A. (1957) Gamma-ray borehole logging. In *Radiometric Methods in the Prospecting of Uranium Ores*, Eds. V. V. Alekseev, A. G. Grammakov, A. I. Nikonov and G. P. Tafeev. Translation available as AEC-tr-3738 (Book 2), US Atomic Energy Agency.
39. Czubek, J. A. (1971) Differential interpretation of gamma-ray logs: I. case of the static gamma-ray curve. Report No. 760/1, Nuclear Energy Information Center, Polish Government Commissioner for Use of Nuclear Energy, Warsaw.
40. Davydov, Y. B. (1970) Odnomernaya obtratnaya zadacha gamma-karotazha skvazhin (One dimensional inversion problem of borehole gamma logging). *Izv. Vyssh. Uchebn. Zaved., Geol. i. Razvedka* **2**, 105–9.
41. Branisa, F. (1974) Filtering of well-log curves. *Geophysics* **39**, 545–9.
42. Conaway, J. G., Bristow, Q. and Killeen, P. G. (1980) Optimization of gamma ray logging techniques for uranium. *Geophysics* **45**, 292–311.

Chapter 4

AUDIOFREQUENCY MAGNETOTELLURIC (AMT) SOUNDING

D. W. STRANGWAY

Department of Geology and Physics, University of Toronto, Ontario, Canada

SUMMARY

This paper describes a number of applications of the audiofrequency magnetotelluric (*AMT*) method. The method uses thunderstorms as a source of electromagnetic energy for probing the earth. When the sources are not too close to the observation site, the source can be considered to be a plane wave. The method described here uses the frequency range from 10 Hz to 10 kHz (which is a useful range for electrical probing) to a few hundred metres in conductive areas, or to several kilometres in resistive terrains.

Because the system has no artificial source, field operations are rapid and simple to carry out. This means that scalar systems can be used to carry out rapid coverage resistivity mapping and hence locate lateral variations in resistivity with a high degree of precision. There are no tensor systems currently in use, although more precise data could in principle be obtained in areas with significant lateral variations in resistivity.

A number of workers are now using controlled sources for AMT (*CSAMT*) observations. It has been shown that if the source is 3–5 skin depths from the observed site, then it can be considered to be a plane wave. The controlled source has the advantage of precision measurements, but the disadvantage of slower coverage. Natural source studies are quite suitable for mapping resistivity where the object is to study large contrasts, and it is unusually suitable for locating boundaries between areas of different resistivity. It is also reasonably precise in areas of uniform stratification, where there is very little lateral variation.

Applications discussed are permafrost studies, massive sulphide exploration, crustal sounding, mapping of resistivities in shales and stratigraphic mapping. Because the source is effectively at infinity, the depth penetration is better than in systems necessarily constrained by the separation between transmitter and receiver.

1. INTRODUCTION

The magnetotelluric method for determining subsurface electrical conductivity structure was first recognised in 1953. Since then most magnetotelluric studies have utilised low frequencies to investigate deep-seated features such as the depth of sedimentary basins or the properties of the earth's crust and mantle. However, Cagniard[1] explicitly outlined the extension of this technique to shallower exploration by using higher frequencies.

During the early summer of 1963, initial field observations were made by the Kennecott Copper Corporation to determine the feasibility of determining ground resistivity by magnetotelluric measurements in the audiofrequency range (from 10 Hz to 10 kHz). The field programme was carried out in conjunction with Deco Electronics (Boulder, Colorado, USA); this company subsequently became a subsidiary of Westinghouse. Sufficient encouragement was realised to justify development of equipment specifically for the purpose of audiofrequency sounding. This original system has been modified and improved so that field measurements can be done simply and routinely. The first detailed published report on the audiomagnetotelluric (AMT) method was by Strangway et al.[2]

Subsequently there have been several published reports describing other systems. Some of these have been described by Hoover et al.,[3,4] and Mabey et al.[5] and Long and Kaufman,[6] who described the results of extensive surveys in the Basin and Range province and in the Columbia river basalts. Ngoc et al.[7] described a system restricted to frequencies of 1, 8, 145 and 3000 Hz, and Benderitter et al.[8] reported on a system operating at 8, 17, 37, 80, 170, 370, 800 and 1700 Hz for a survey in Finland. Dupis and Iliceto[9] and Dupis et al.[10] described applications of this system to the Lardarello geothermal site. Slankis et al.[11] reported on a system that operated at 8 Hz and Guineau[12] reported on a system operating between 10 kHz and 500 kHz at much higher frequencies than those considered here.

The first conference on this subject was organised at the 1981 meeting of the International Association of Geomagnetism and Aeronomy in Edinburgh, UK. At this meeting both natural source and controlled source work was described. Natural source work was described by groups from the University of Toronto, Canada; University of Edinburgh, UK; Observatoire Cantonal in Neuchatel, Switzerland; University of Oulu, Finland; Mining Companies in Norway and Finland; and Macquarie University in Australia.

The first applications of artificial source (AMT) were described by Goldstein[13] and by Goldstein and Strangway.[14] Subsequent published work has been limited, but there has been extensive field work using this method as indicated in the reports given at the Edinburgh meeting by Van Blaricom as well as by Mussman and Otten and by Lakanen. An excellent report was given by Sandberg and Hohmann[15] describing an example of controlled source AMT (CSAMT).

It is clear that audiofrequency magnetotellurics is a method which has been gaining increasing acceptance because of its simplicity, its ability to carry out surveys rapidly, and the depth penetration potential because the source is at a large distance. The method is particularly useful when operated as a mapping device, since it has the capacity for very high lateral resolution. Because the source is effectively at infinity, it has the capacity for greater depth penetration than any comparable electromagnetic method which is limited by the distance between transmitter and receiver.

2. METHODS

2.1. Principle of the Method

The magnetotelluric method has been described by Cagniard,[1] Wait,[16] Vozoff,[17] and others. Basically, the electromagnetic impedance—the ratio of the horizontal electrical field (E) in the ground to the orthogonal horizontal magnetic field (H)—is measured at a number of frequencies to yield earth resistivities as a function of frequency (resulting in a form of depth sounding). Operationally, the technique is quite simple, in that depth sounding is conducted by making measurements at only one location.

The natural magnetotelluric source fields can be described either in terms of a propagating electromagnetic wave interacting with the earth's surface, or as an inductive phenomenon in which telluric currents are induced by the fluctuating geomagnetic field. The incident, horizontal magnetic field, is roughly doubled at the surface of the earth relative to the cavity, and it is

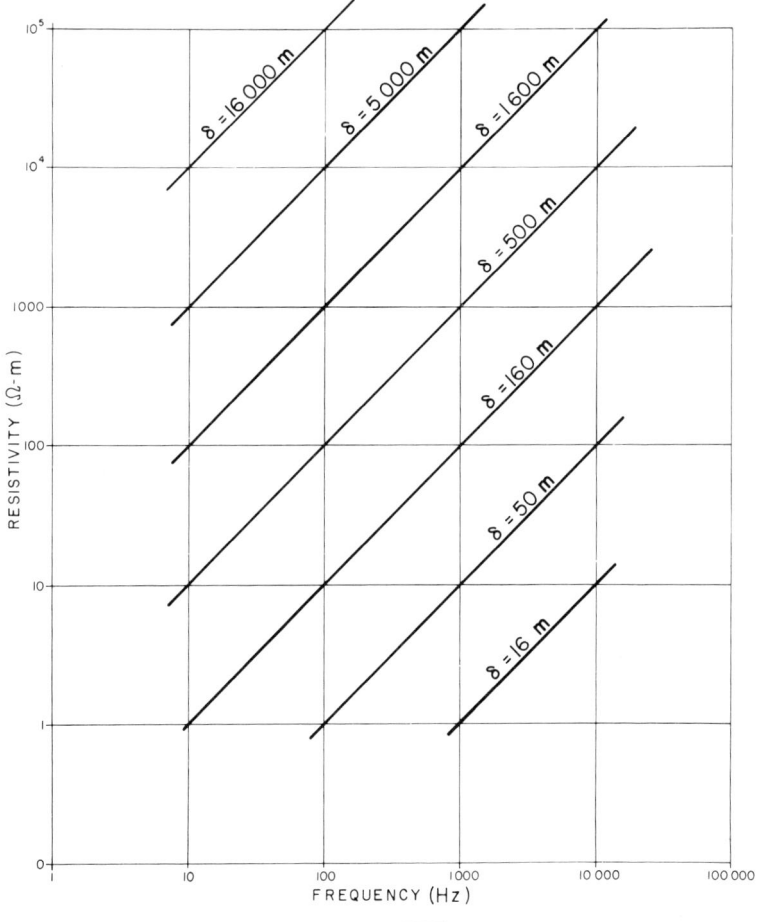

SKIN DEPTH $\delta = \sqrt{\frac{2}{\mu\omega\sigma}}$, in meters

FIG. 1. Skin depth plot for various combinations of resistivity and frequency (assumimg a unit value of the magnetic permeability).

uniform in the absence of lateral contrasts in resistivity. The electric field, on the other hand, is directly dependent upon the earth's resistivity structure.

The depth of sounding can be roughly related to frequency by use of the skin depth, defined as

$$\delta = \sqrt{\frac{2\rho}{\mu\omega}}$$

AUDIOFREQUENCY MAGNETOTELLURIC (AMT) SOUNDING

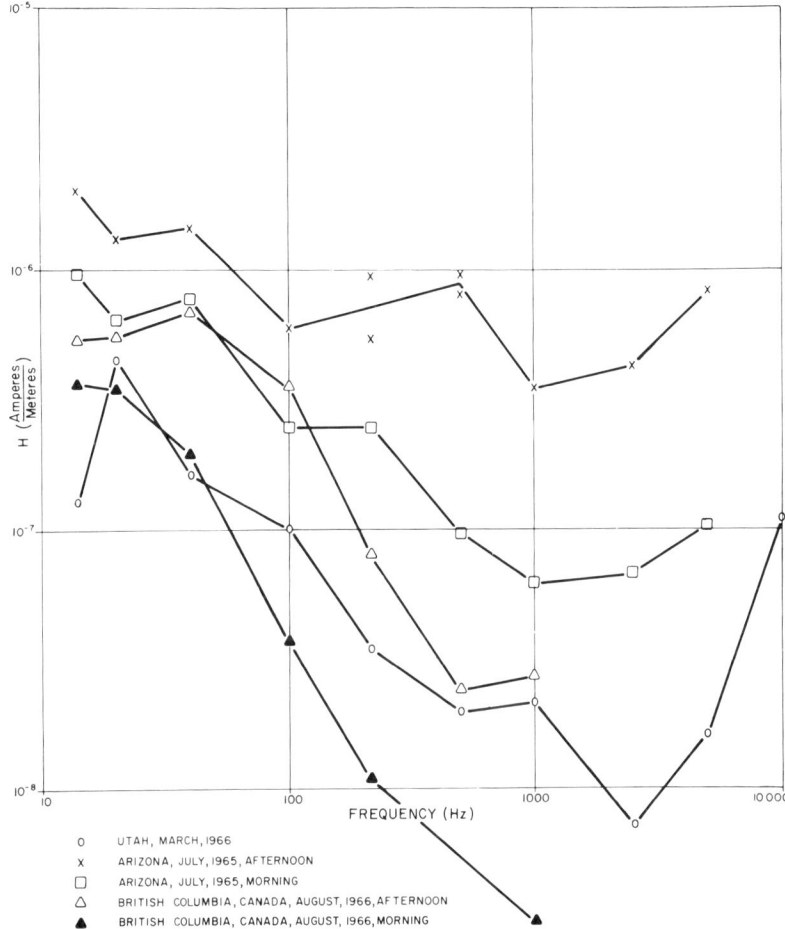

FIG. 2. Field strength plot (amp. turn/metre) in the audiofrequency range taken at various times. This illustrates the 2 kHz null when sources are distant.[2]

where δ is the skin depth in metres, f is the frequency in hertz ($\omega = 2\pi f$), ρ is the resistivity in ohm metres and μ is the magnetic permeability in henry/metre. Plots of skin depth as a function of resistivity and frequency are shown in Fig. 1. In this figure it can be seen that the audiofrequency range from 10 Hz to 10 kHz, covers the depth range of a few metres to a few kilometres, depending on the resistivity. The method can thus be viewed as an alternative to other more conventional methods, for determining near-surface resistivities.

The apparent resistivity (ρ_a) can be determined simply from the measured impedance (E/H) by

$$\rho_a = \frac{1\cdot 26 \times 10^5 \times (E/H)^2}{f}$$

where E is in volt/metre, and H is in amp.turn/metre. The measurements can then be displayed on a log resistivity versus log frequency plot to give the standard magnetotelluric sounding curve at each station occupied.

2.2. Source Fields

The problem of source field in the audiofrequency range has been considered by many workers in studies of radio propagation. Interest in natural noise fields is considerable, since they can interfere with communication systems. The Afmag technique commonly used in geophysical exploration also makes use of natural audiofrequency signals at 150 Hz and 500 Hz. It is well known and understood that the main source of natural noise in the frequency range of 10 Hz–10 kHz is thunderstorms, which are extensively concentrated in the tropics and tend to peak in their activity in the early afternoon, local time (Fig. 2). The generated electromagnetic energy then propagates around the world, trapped in the waveguide formed between the earth's surface and the ionosphere (Budden[46]).

2.3. Interpretation

The simplest method of interpreting magnetotelluric data is the comparison of the sounding curves obtained in the field with theoretical curves. Several workers have published theoretical apparent resistivity versus frequency curves for two- and three-layer cases, and the calculation for multi-layer configuration are given in Fig. 3. Because of the law of electromagnetic similitude, these layered-media curves can be used for any combination of layer thickness and resistivity. Specifically, if we consider any configuration, and multiply all lengths by K_L and all resistivities by K_ρ, the original curve is correct if all frequencies are multiplied by the factor

$$K_F = \frac{K_L}{K_\rho^2}$$

This operation essentially keeps constant the electromagnetic response parameter kd (or $k^2 d^2 = i\mu\omega/\rho$, where d is a representative length).

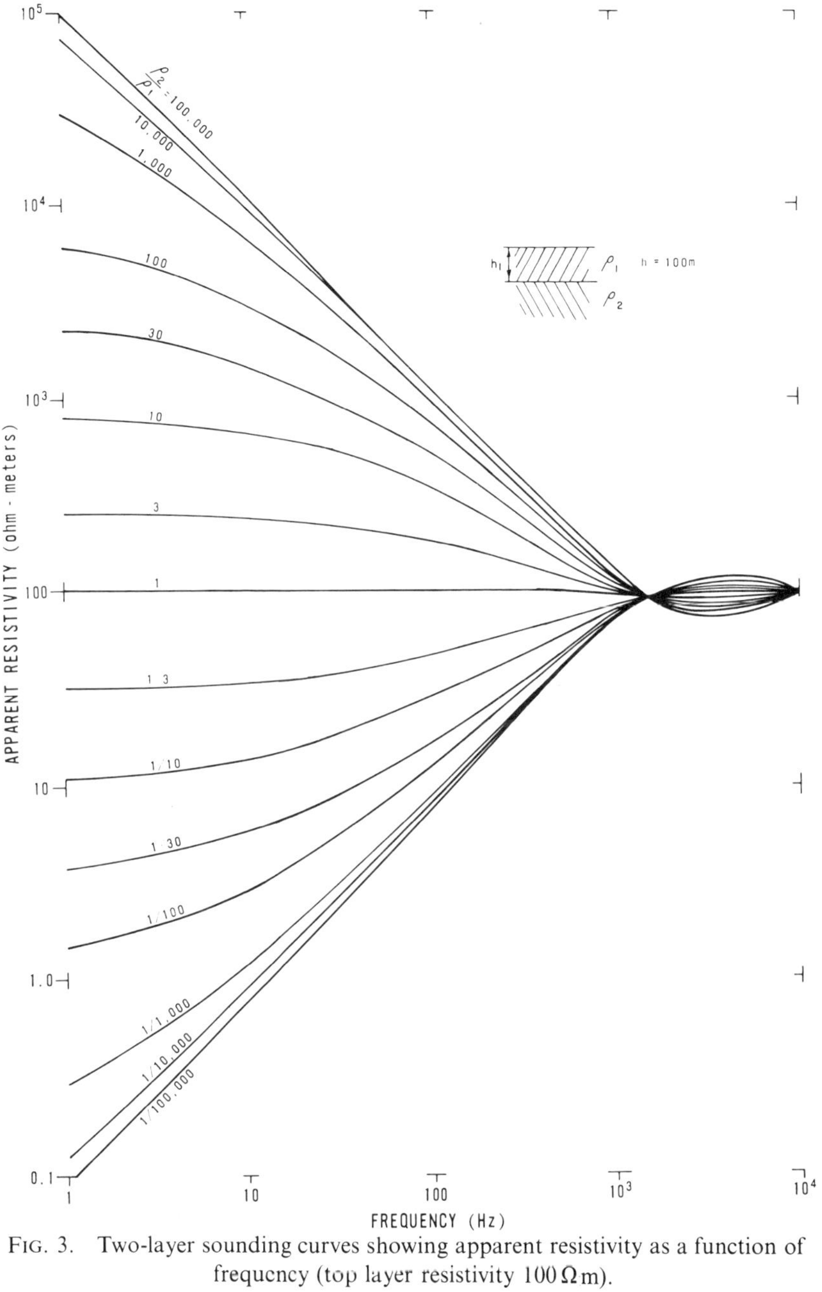

AUDIOFREQUENCY MAGNETOTELLURIC (AMT) SOUNDING

FIG. 3. Two-layer sounding curves showing apparent resistivity as a function of frequency (top layer resistivity $100\,\Omega\,\text{m}$).

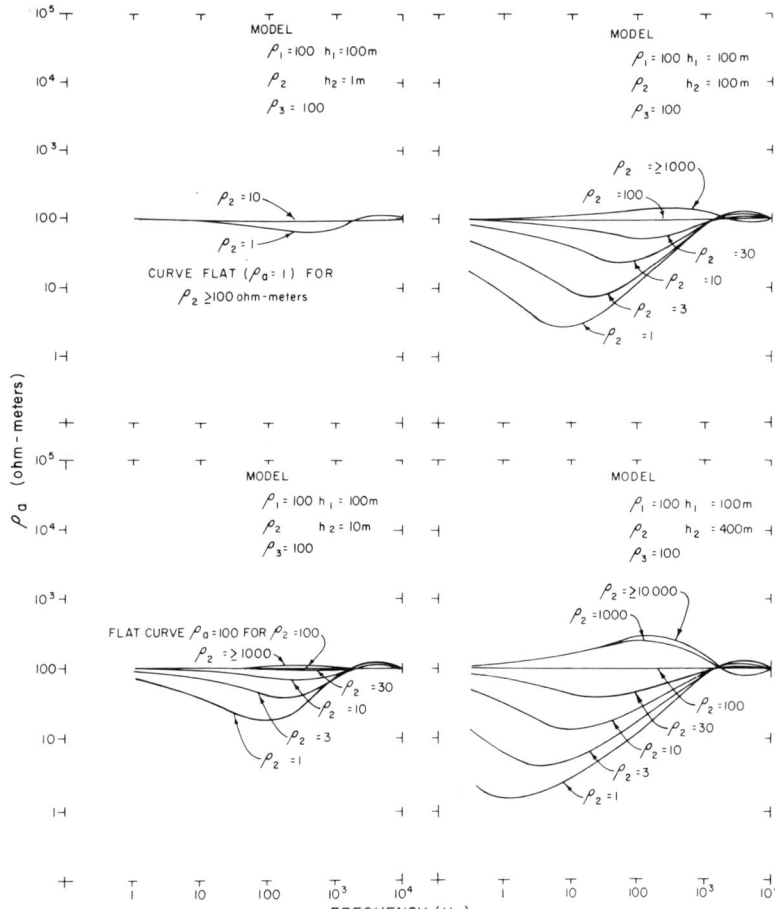

FIG. 4. Three-layer sounding curves in which the top and bottom layers have the same resistivity (100 Ω.m) showing the resolution of the middle layer for varying resistivity and thickness.

Theoretical curves to show the effect of a layer sandwiched between an upper layer and a lower half-space are given in Fig. 4. These curves are similar to those of Yungul.[18] Essentially, it is not possible to detect a conducting middle layer which is 1/100 the thickness of the upper layer. On the other hand, if the thickness of the middle layer is as little as 1/30 that of the upper layer and it is a very good conductor, clear indications should be present. Layers of high resistivity, however, are hard to detect and it is not until the thickness of the middle layer is 2 or 3 times that of the upper layer that it becomes detectable as a separate layer. Limitations in measurements thus lead to the fact that true and accurate depth sounding is very difficult. The method, as other inductive techniques, is particularly suited to searching for low-resistivity zones.[19,20]

Frequently, however, the conductivity structure of the earth cannot be accurately represented by a layered-media approximation. For instance, in mineral exploration, a prime geophysical objective is often delineating the lateral extent of an anomalous conductive zone marked by a vertical contact.

Fundamentally, electrical current tends to flow into a medium of high conductivity. Therefore, current induced by the geomagnetic field may flow in a direction controlled by the local geology rather than in a perpendicular direction to the magnetic field source component, as expected when no lateral resistivity contrast is present. Moreover, lateral resistivity variations result in adding a vertical magnetic field component. This is the component used in the Afmag technique. Therefore, field complications arise because the measured impedance at the surface of the earth is no longer independent of either the orientation of the orthogonal fields measured or the polarisation of the incoming source field. As a result, the measured scalar apparent resistivity in the immediate vicinity of lateral resistivity contrasts can have very poor repeatability in time, especially over periods of hours when the source field thunderstorms may have changed positions. Similarly, two measurements using field recording orientations rotated 90 degrees can show drastic differences in the observed apparent resistivity curves. It is for this reason that tensor systems would be of interest. No tensor systems have yet been developed for the audiofrequency range although there are several groups working on such a system.

Because the magnetotelluric impedance is a tensor quantity for an earth with a two-dimensional resistivity structure, simultaneous phase-sensitive measurement of both orthogonal components of the electric and magnetic fields is required for a complete description. Broadband, four-channel recording and subsequent Fourier analysis is commonly utilised in low-frequency magnetotelluric studies.[21-23] The AMT method as presently practised however is largely useful as a rapid and cheap system which depends on its rapid coverage and field-portability for its effectiveness. Systems designed to measure only scalar resistivities are thus in normal use, but strictly valid data are obtained only when the measuring orientation is aligned with the geologic structure. By making many closely-spaced readings, which is possible with a light, simple, portable system, however, the position of lateral contacts can be exceedingly well defined, and those regions where anisotropies exist can be clearly delineated.

The results of many field studies show that it is quite possible to locate sharp lateral contrasts, such as faults with considerable precision and hence to locate regions of stratified resistivity for interpretation. It is quite frequently the case that repeatability is poor near such contacts. On the other hand, in areas of uniform layering, repeatability is frequently very high and scalar data can be used to give precise depth sounding information and useful inversions.

With the AMT apparatus aligned with the structure, it is possible to directly measure the impedance for 'E-parallel' or 'E-perpendicular' polarisations (the electric field detector oriented parallel to or across the structure, respectively). Apparent resistivities obtained from these impedances can then be compared with theoretical two-dimensional-type curves corresponding to the appropriate source polarisation. Theoretical, apparent resistivities for arbitrary two-dimensional resistivity structures can be calculated by use of computer modelling techniques. This procedure recognises the analogy between Maxwell's equations for a two-dimensional geometry and the transmission line, and constructs an electrical network in which the circuit elements depend upon the frequency and resistivity structure and the voltage and current relate to the electric and magnetic field values. The technique is discussed by Madden and Swift,[19] and is elaborated upon by Swift.[24]

2.4. Instrumentation

The instrumentation used in such surveys consists of two sensors (electric and magnetic field), a prewhitening filter and a variable frequency analyser. The electric field sensor is simply a long cable grounded at both ends. Typical experiments use 50 m of wire for the detector, although it is possible to use almost any convenient length.

Magnetic field sensors are usually ferrite-cored coils such as one developed at the University of Toronto. It has a response that is linear with frequency from less than 10 Hz to 40 kHz. Preconditioning circuitry is necessary to provide rejection of 60 Hz (50 Hz in Europe) and both odd and even harmonics, and to prewhiten the signal before presentation to the frequency analyser. This also consists of band-limiting the signals, since large signals from VLF stations and broadcast stations can saturate the system while attempting to detect very small signals at the frequency of interest.

Analysers can be of many different kinds. The one in use by our group contains a central oscillator and a heterodyning system to measure the ratio of the two signals (E and H) fed to it. The filtering characteristics of the system must be extremely sharp and provide narrow-band detection at the frequency of interest. In addition, the skirts must be very sharp, since the system is required to detect very small signals in the presence of very large ones at nearby frequencies. Skirt rejections of 10^5 or more are required in order to reduce aliasing by strong power-line related frequencies.

The rest of this paper will be devoted to a series of examples which describe how the AMT method has been applied in a variety of geologic circumstances. The first category describes briefly a series of cases in which the geologic features are essentially horizontally layered. This means that they may be considered as one-dimensional for interpretation purposes. The fact that these can be viewed as one-dimensional cases can be readily tested by using a profile or grid of sounding stations showing that there are only minor lateral variations.

Since the skin depth in materials of low resistivity is small, it is clear that the chances of finding one-dimensional cases is considerably enhanced in low resistivity environments, while in resistive environments the sounding involves much larger volumes of rock, and hence there is a much greater chance of encountering lateral variations. An example of one-dimensional studies not reported here is contained in papers by Strangway et al.[25] and by Ilkisik et al.[26] in which sounding through conductive glacial clays has been quite successful.

3. RESULTS

3.1. One-Dimensional Examples
3.1.1. Southern Ontario—Palaeozoic Sediments
An experiment was done near the town of Milton, west of the Niagara escarpment, about 50 km west of Toronto. The geology and location of the test site are shown in Fig. 5. The geology of the Palaeozoic sedimentary basin of southern Ontario is well documented, and other measurements using different electromagnetic techniques have been reported.[27]

The AMT sounding data obtained at Milton are shown in Fig. 6(a). At first glance the data appear to suggest a two-layer earth with a more resistive layer at surface overlying a more conductive layer at depth. The stratigraphic sequence in this area consists of interbedded shales and limestones with some evaporites. Some reef structures have been found in this area. A major conductor in the section corresponds to a layer of shale,

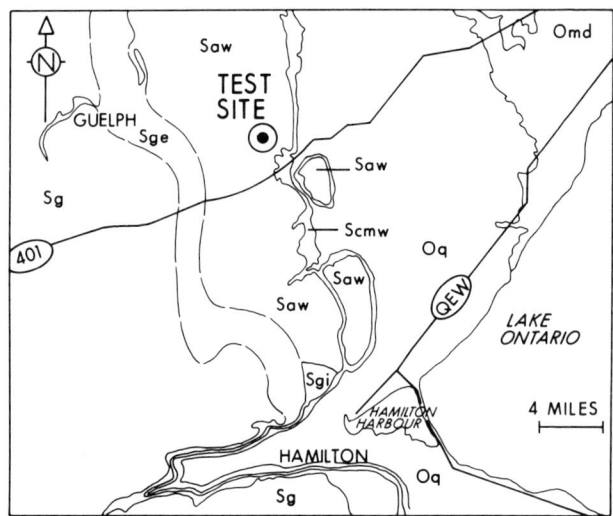

FIG. 5. Location and geologic map of the Milton site sounding in southern Ontario. Units labelled S are Silurian and those labelled O are Ordovician. Sg, Guelph Form; Saw, Wiarton Form; Sgi, Goat Island Form; Scmw, Cataract Gp; Oq, Queenston Form; Omd, Meaford-Dundas Form; QEW, Queen Elizabeth Way.

about 400 m thick, as determined by drilling (Toronto–Windsor Geological Map)[28] and illustrated in Fig. 7. The conductive layer is sandwiched between a more resistive dolomite (about 30–40 m thick) at the surface and a layer of limestone at a depth of 430 m. The sediments overlie the Precambrian basement at a depth of 650 m, which is expected to be highly resistive.

Figure 6(a) shows an initial guess corresponding to a four-layer model. After inversion and reduction to the best least-squares fit using the Marquardt method employed by Hsu,[29] the model shown in Fig. 6(b) is determined. This model corresponds reasonably well with the drilling results. A layer of material (280 Ω m), about 40 m thick, overlies a more conductive layer (23 Ω m) about 75 m thick, which in turn overlies a layer about 120 m thick of 11 Ω m materials. It is a total of 235 m to the base of this layer. The lowest layer is also conductive and probably represents more conductive shales in the stratigraphic sequence of Fig. 7. The maximum penetration depth is greater than 235 m, but neither the limestone below 430 m or the basement were penetrated. It seems clear that the method has detected clear stratigraphic layering in the resistivity structure. Because of

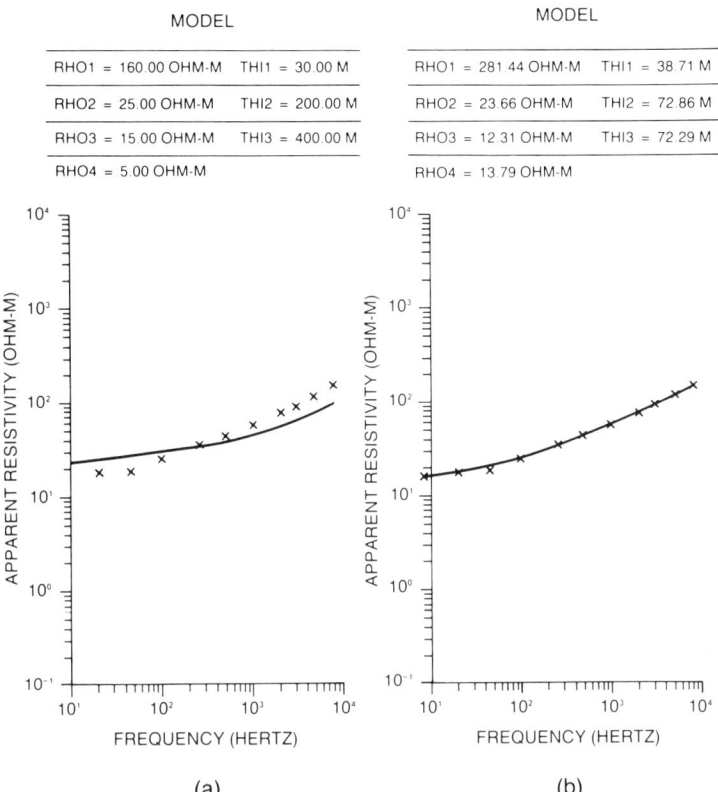

FIG. 6. Plot of sounding data from Milton, × show observations of apparent resistivity at frequency shown. The line represents the four-layer curve corresponding to the model shown at the top. (a) First approximation used to initiate the inversion; (b) final model derived using the Marquardt least squares approach of Silvester and Haslam;[42] from Hsu.[29]

the conductive shales however, it was not possible to detect either the basal limestone or the Precambrian basement.

In Fig. 8, we show the results of modelling by Duncan et al.,[27] who used a line source method with frequencies as low as 1 Hz to carry out electrical sounding in this same area. In the case, by using the lower frequencies, a more resistive layer was detected at a depth of 470–500 m. This layer was not detected in our study and appears to correspond to the basal limestone. It can be seen that the AMT method gives similar results to those determined using other electrical methods, thus confirming that the AMT

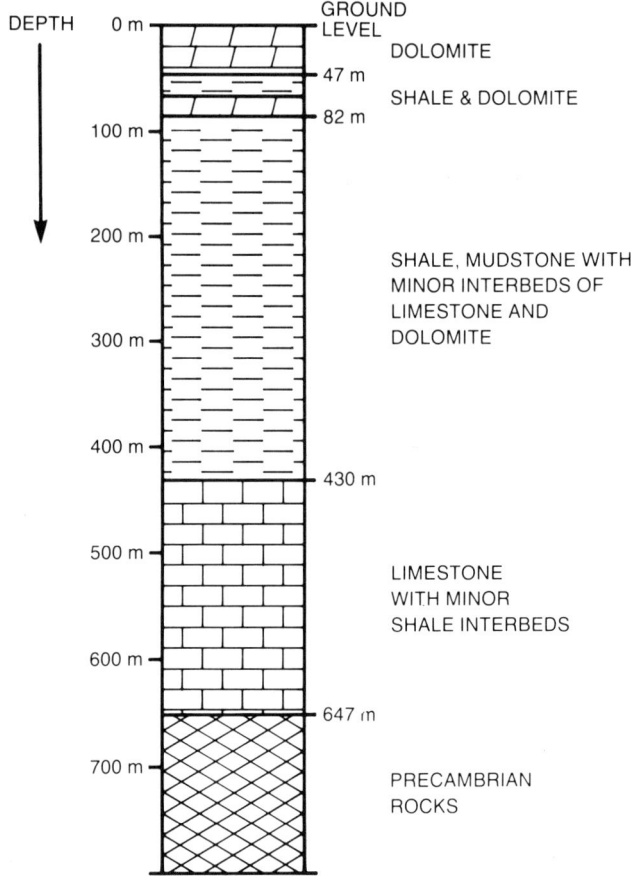

FIG. 7. Geologic cross-section at the Milton test site based on drilling results.

method can give rapid and useful information on layered stratigraphic sequences.

3.1.2. Williston Basin—Manitoba and Saskatchewan

Figure 9 gives the location of two sets of soundings carried out in the Williston basin in south-eastern Saskatchewan and south-western Manitoba.[30] In each case there were a series of stations positioned along profiles. The results of these soundings showed a remarkable uniformity from station to station, confirming that the sediments were laterally uniform in properties and uniformly layered. Pseudosections from the

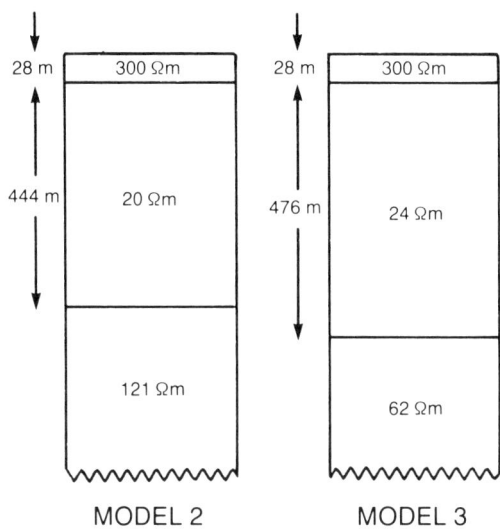

FIG. 8. Models from Duncan et al.[27] based on inversion of electromagnetic sounding data at the Milton site based on a long-wire source method. Model 2 is based on amplitude data and Model 3 on phase data.

Dumas area (Saskatchewan) and the Hartney area (Manitoba) are shown in Fig. 10, clearly illustrating that in the frequency range from 10 Hz to 10 kHz, the data suggest a simple two-layer earth. A layer with a resistivity of about $10-12\,\Omega\,m$ is at the top, overlying extremely conductive material of only $2-3\,\Omega\,m$. These sediments are so conductive that penetration depths are extremely limited, even at the lower frequencies. The surface sediments are Cretaceous shales and we infer that this is the cause of the very low resistivities. The surface layer is about 40–50 m thick and we infer that this corresponds roughly to the overburden in this region. It is clear that the AMT method can only sound effectively to depths of 50–100 m in this environment. The main application of the method in such environments is to determine the thickness of the upper layer and to map possible variations in the resistivity of the shales along traverses.

Independent studies were conducted by Rankin and Kao[31] at much lower frequencies covering the range from 5×10^{-4} Hz to 8 Hz. These soundings were carried out in the Williston basin very close to the Hartney and the Dumas sites. This means that a range of frequencies from 5×10^{-4} Hz to 10^4 Hz was available for interpretation. For Dumas, the sounding data and a least-squares model fit are shown in Fig. 11. This represents a residual least-squares estimation for a seven-layer earth. Both

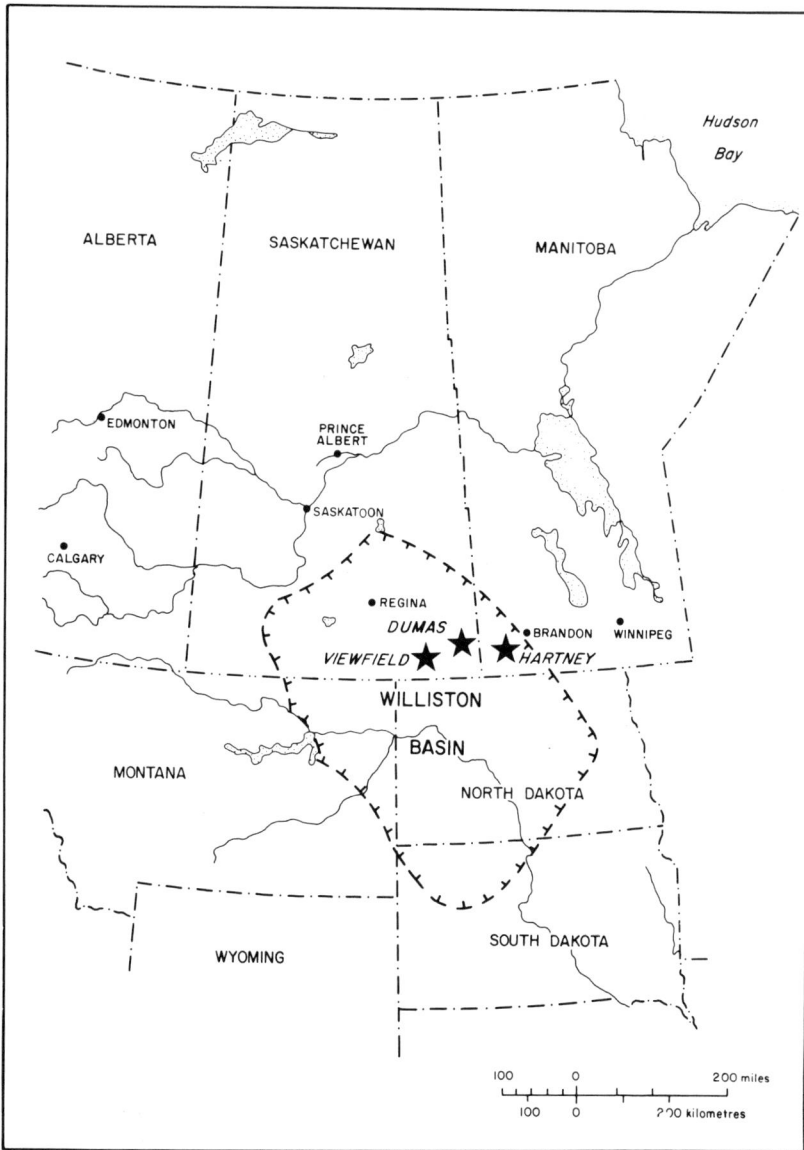

Fig. 9(a). Location map showing the location of the AMT sounding sites of Viewfield, Dumas and Hartney.

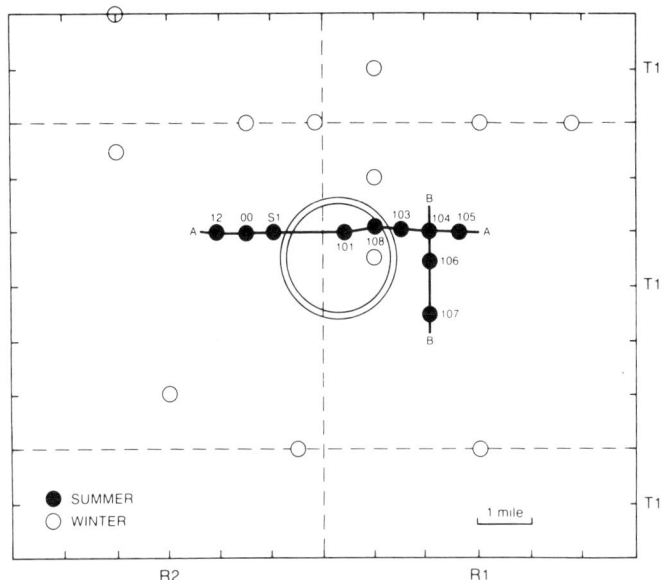

FIG. 9(b). Stations observed at the Dumas site (T11, R1 and 2, W2M). Solid circles were observed both in summer and in winter. Profile data shown in Figs 10(a) and (b) are along line A. The large circle is the location of a collapse structure at depth.

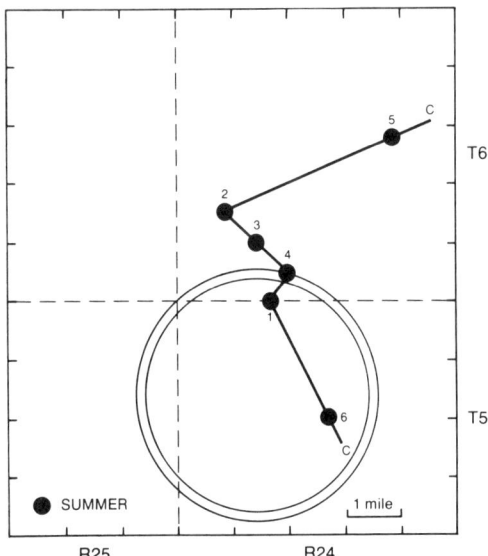

FIG. 9(c). Location of stations at the Hartney site in Manitoba (T5, R24, W1M). Data shown in Figs 10(c) and (d) are along profile C. The large circle is the location of a collapse structure at depth.

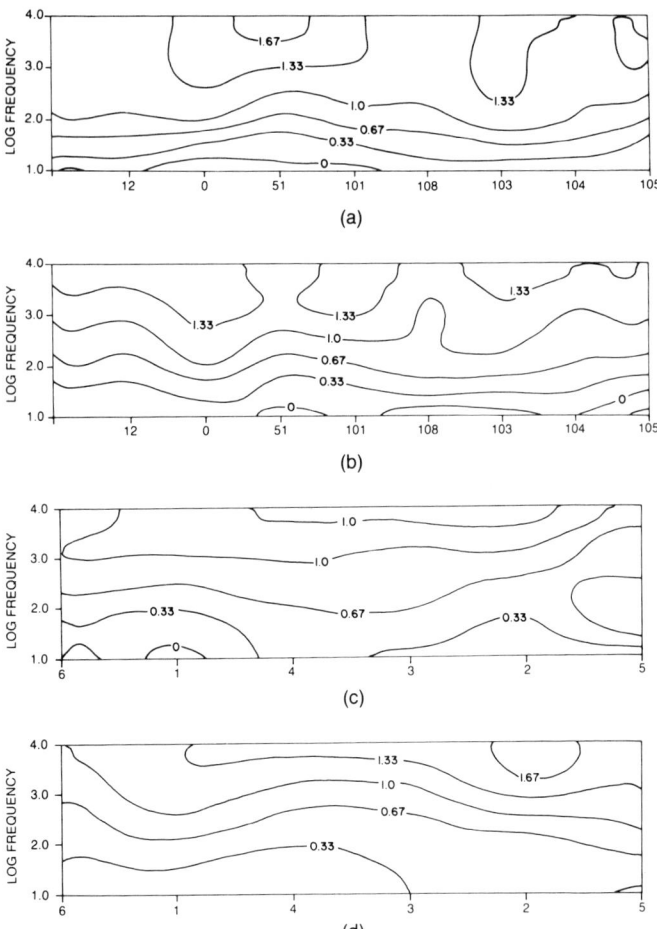

FIG. 10. Pseudosections along profiles A and C of Fig. 9. Pseudosections are contours of log ρ_a in units of 3 intervals per decade. The horizontal axis is positioned along the line with the stations shown. The vertical axis is the apparent resistivity in a logarithmic scale. The contours are derived in several steps: (i) a third-order polynomial fit is made on each sounding at each station; (ii) this fit is then interpolated between each station using ρ_a and (iii) the contouring is done. (a) Dumas area line A—electric dipole oriented north–south; (b) Dumas area line A—electric dipole oriented east–west; (c) Hartney area line C—electric dipole oriented north–south; (d) Hartney area line C—electric dipole oriented east–west. It is noteworthy that the contours are nearly horizontal, suggesting horizontal layering, and that orthogonal measurements give very similar results. The resistivity is higher at high frequencies and drops with frequency, i.e. with depth.

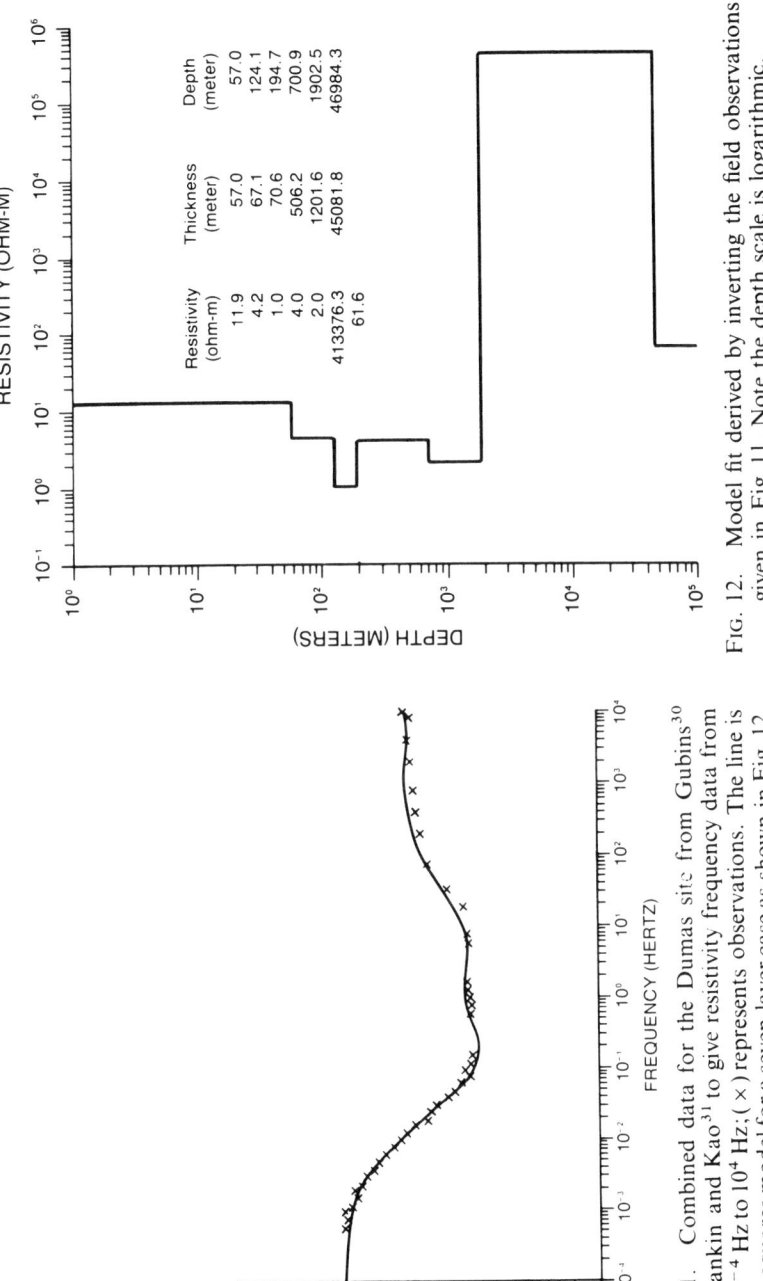

FIG. 12. Model fit derived by inverting the field observations given in Fig. 11. Note the depth scale is logarithmic.

FIG. 11. Combined data for the Dumas site from Gubins[30] and Rankin and Kao[31] to give resistivity frequency data from 5×10^{-4} Hz to 10^4 Hz; (×) represents observations. The line is a least-squares model for a seven-layer case as shown in Fig. 12.

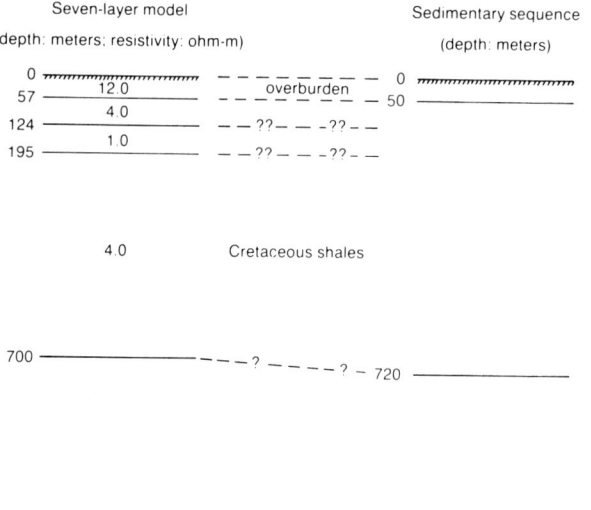

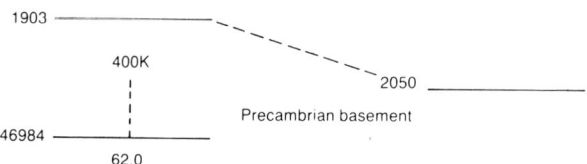

FIG. 13. Comparison between the inverted data of Figs 11 and 12 and the known stratigraphic section in the Dumas area. Note the good determination of the depth to basement and the shallow stratigraphic layers inferred.

the model fit and the field data are shown in Fig. 11. This fit corresponds to the model shown in Fig. 12, which shows the inferred resistivity (log scale) versus depth (log scale). The surface layer is somewhat more resistive than the rest of the sediments, where the value drops to a low of 1 Ω m. At a depth of 1900 m, the resistivity appears to rise sharply to a very large value. The contrast is sufficiently large that the value is not well-determined. We know only that it is several tens of thousands of ohm metres in order to give such a steep rise in the curve. At the lowest frequencies, the value levels off rapidly and it is logical to infer that at a depth of 40 km or more there is a sharp drop in resistivity. The results of this inversion are compared in Fig. 13, with drilling results from the area. There is a general correspondence between the inverted results and depth to basement. There is perhaps even a correlation between the resistivities and a stratigraphic break at 700 m.

A similar set of results was obtained in the Hartney area as shown in Figs 14, 15 and 16. The audiofrequency data were measured using both north–south and east–west data, and no significant differences were observed, suggesting that the material is isotropic and laterally homogeneous. Rankin and Kao's data[31] at low frequencies, however, shows strong anisotropy. Since their data was based on a single observation, we have no way of knowing the cause of this feature. It seems most likely that there are strong conductors in the basement of the type described by Strangway *et al.*[25]

Nevertheless it was possible to derive the model shown in Figs 15 and 16 in which the resistive basement is inferred to be at a depth of about 1450 m. There appears to be a higher resistivity section of dolomite and limestone (29 Ω m) and a sharply defined conductive shale horizon at the basement interface. Again it appears to be 40 km or so to the conductive upper mantle.

In the sedimentary basin of western Canada, the AMT method appears to be a useful rapid mapping tool for detecting small resistivity changes in the conductive shales as well as mapping a resistive (overburden and/or weathering-related) surface layer. Penetration depth is only a few hundred metres, but it is a useful addition to information used for deep crustal sounding using lower frequencies.

3.1.3. Athabasca Sandstone (Saskatchewan)

We now turn to a very different example of the use of the AMT method in a one-dimensional study. In northern Saskatchewan, the Athabasca sandstone of late Precambrian age, unconformably overlies the Archean

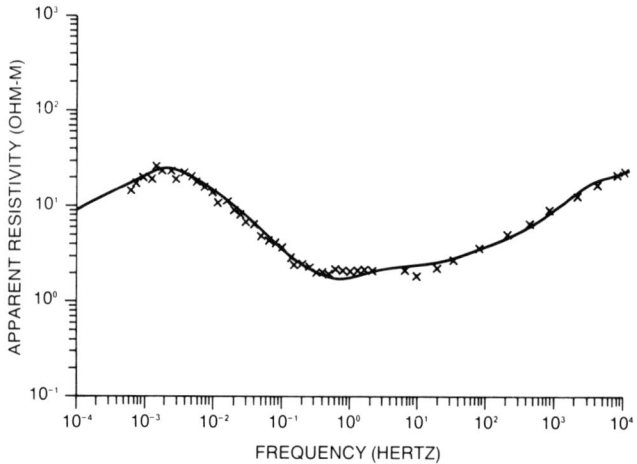

FIG. 14. Combined high and low frequency data from Hartney, Manitoba showing observation points (×) and seven-layer model fit.

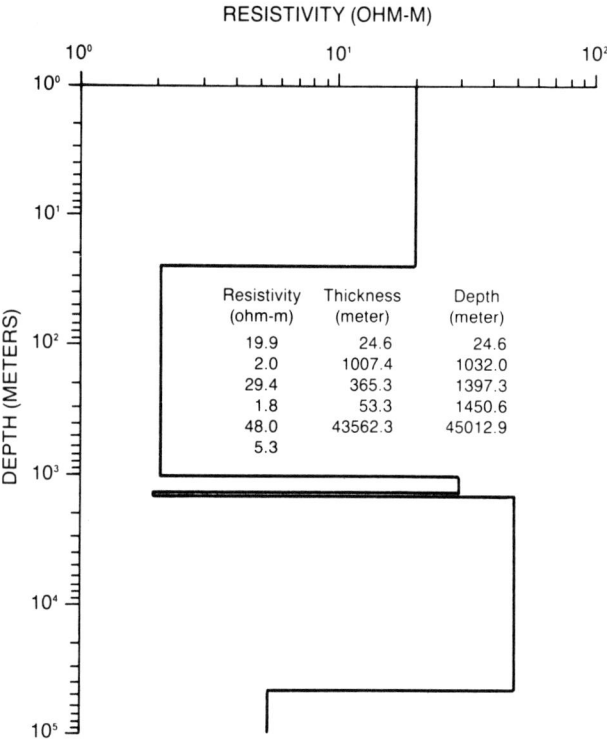

Resistivity (ohm-m)	Thickness (meter)	Depth (meter)
19.9	24.6	24.6
2.0	1007.4	1032.0
29.4	365.3	1397.3
1.8	53.3	1450.6
48.0	43562.3	45012.9
5.3		

FIG. 15. Model derived from inverting the field observations of Fig. 14.

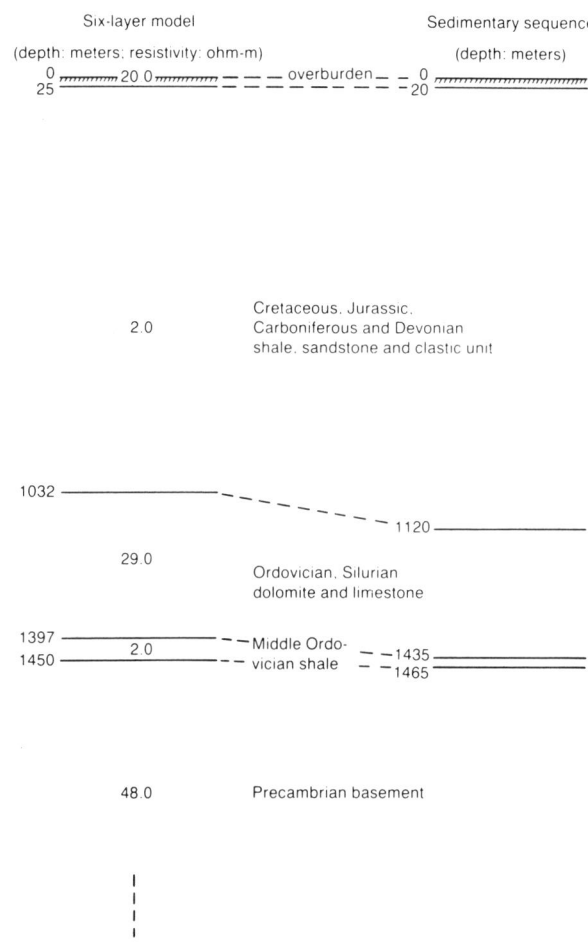

FIG. 16. Comparison of inverted magnetotelluric data and known stratigraphic section at Hartney, Manitoba.

130   D. W. STRANGWAY

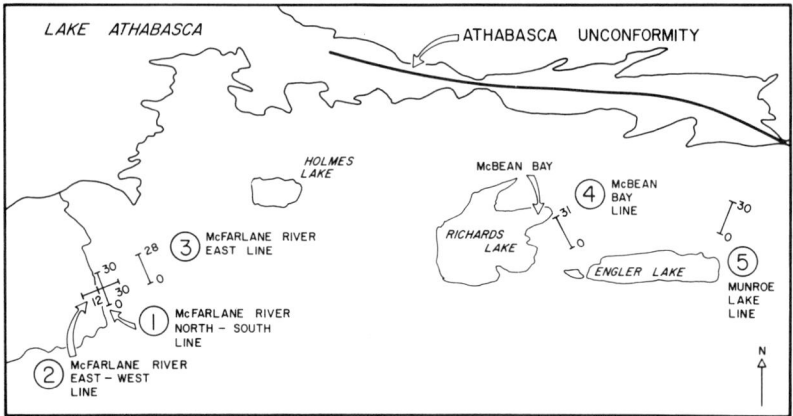

FIG. 17. Location map in northern Saskatchewan showing sounding profiles in the Athabasca sandstone area south of Lake Athabasca.

basement. This sandstone is very thick in some areas, but in general it is very uniform in its character. The sandstone is somewhat metamorphosed and as a result it has a high resistivity when compared to younger Mesozoic or Palaeozoic sediments.

A location map for a survey in this area is shown in Fig. 17. The data that are discussed here consist of a set of profiles from the northern part of the

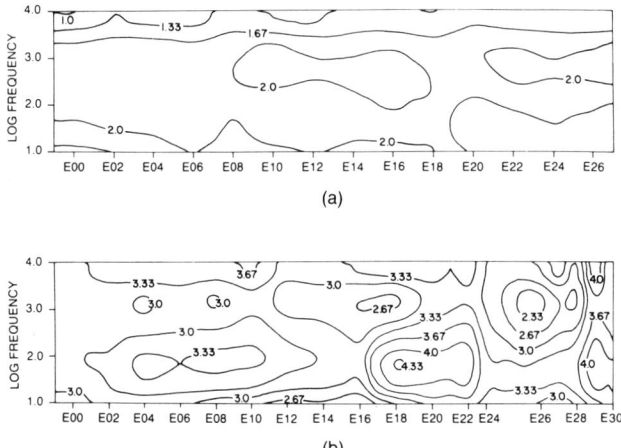

FIG. 18. (a) Pseudosections from line 3 from one of the western sites shown in Fig. 17. (b) Pseudosections from line 4 in the eastern part of the area.

sandstone, just south of Lake Athabasca and involved a total of about 100 stations. Representative pseudosections from the eastern and the western areas are shown in Fig. 18. The most remarkable feature of these profiles is the uniformity along the profiles. These two areas, while different from each other, are remarkably uniformly layered and sounding and one-dimensional inversions can be derived with good confidence in the results.

The two areas examined are quite different. In the western area, the results can be readily fitted to a four-layer earth as shown in Fig. 19. This shows that the overburden has a resistivity of about $6000\,\Omega\,\text{m}$ and a thickness of 70 m. The sandstone has a resistivity of about $250\,\Omega\,\text{m}$ and has a thin conductive layer at the top. The depth to the base of the $250\,\Omega\,\text{m}$ sandstone is about 1000 m and then a conductive layer is present. This thickness is typical for the sandstone which commonly has a weathered regolith layer at the base.

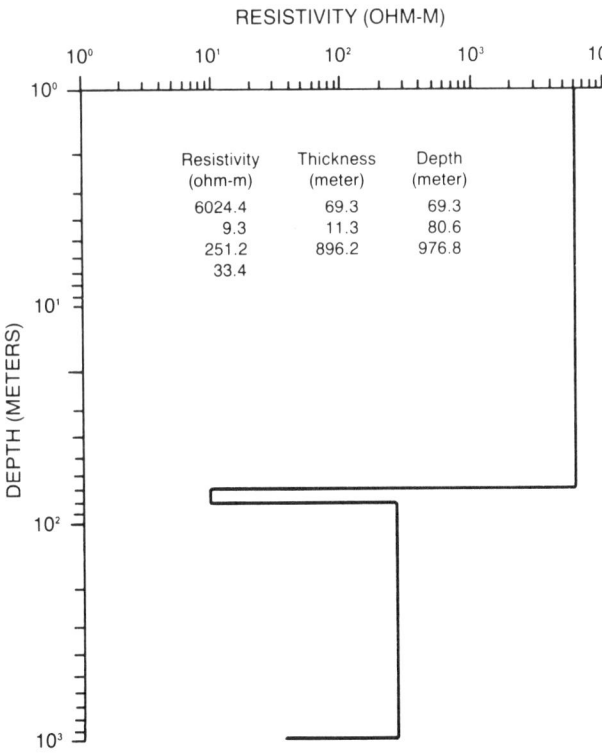

FIG. 19(a). Four-layer model fit which best fits the data averaged along line 3 in the western part of the region.

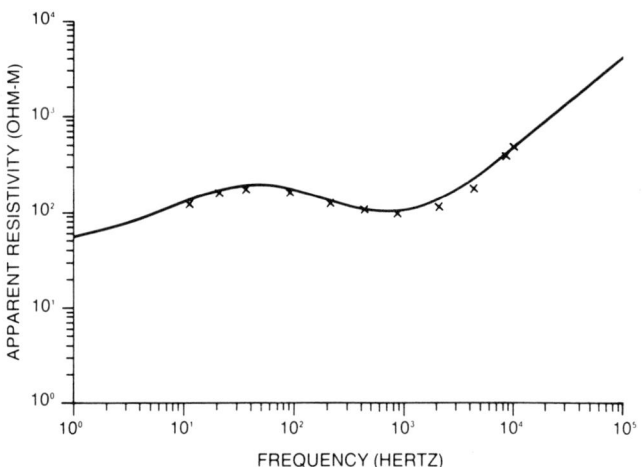

FIG. 19(b). Comparison between observations (×) and the model results obtained with the layers shown in (a).

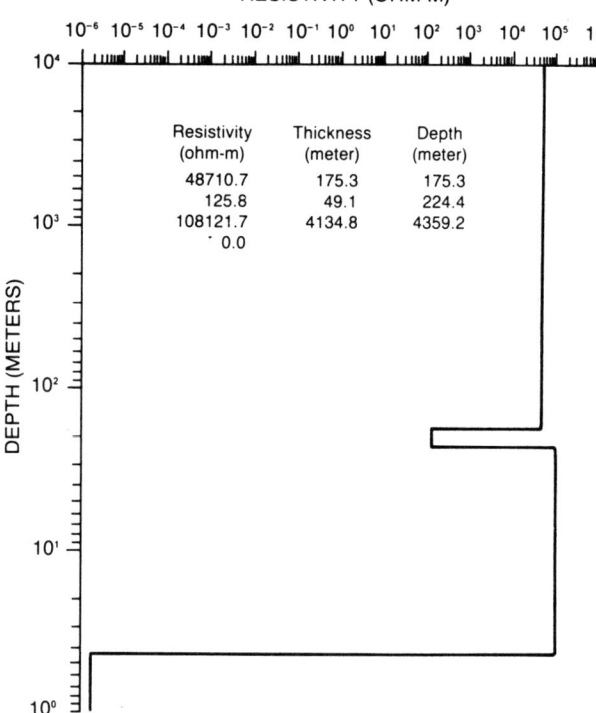

FIG. 20(a). Four-layer model fit which best fits the data averaged along line 4 in the eastern part of the region.

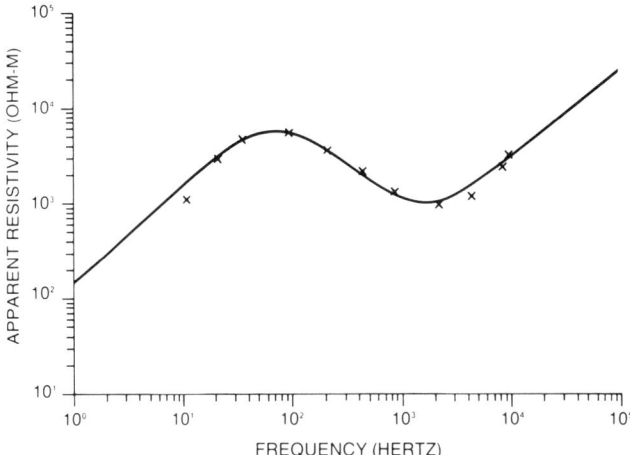

FIG. 20(b). Comparison between observations (×) and model results obtained with the layers shown in (a).

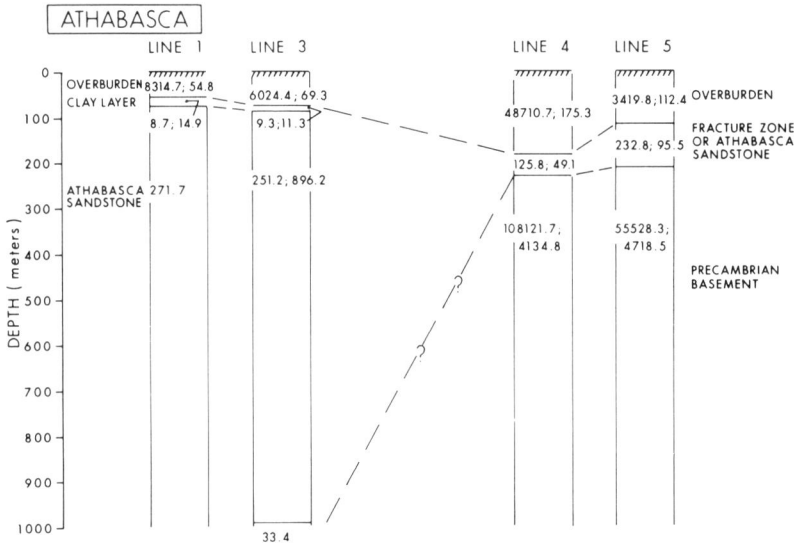

FIG. 21. Inferred geologic cross-sections based on the four-layer inversions for lines 1, 3, 4 and 5. Numbers in each layer are first the resistivities (Ωm) and then the thicknesses (m). Note the similarities between lines 1 and 3 and between lines 4 and 5 and the differences between the two sets.

In the eastern part of the area, the sounding results are much more like those of typical Archean basement rocks[25,32] as shown in Fig. 20. We can only conclude from this that the sandstone is very thin or extremely resistive. A schematic profile showing the results of inversion of data on lines 1, 3, 4 and 5 is shown in Fig. 21. These results are representative of each profile and suggest that this is a rapid effective mapping tool with a considerable depth penetration capacity in uniformly layered, resistive environments.

3.1.4. Mesozoic/Palaeozoic Sediments, New Mexico

A detailed survey in the vicinity of Zuni, New Mexico, was carried out in connection with a geothermal prospect (Fig. 22). A total of 119 stations

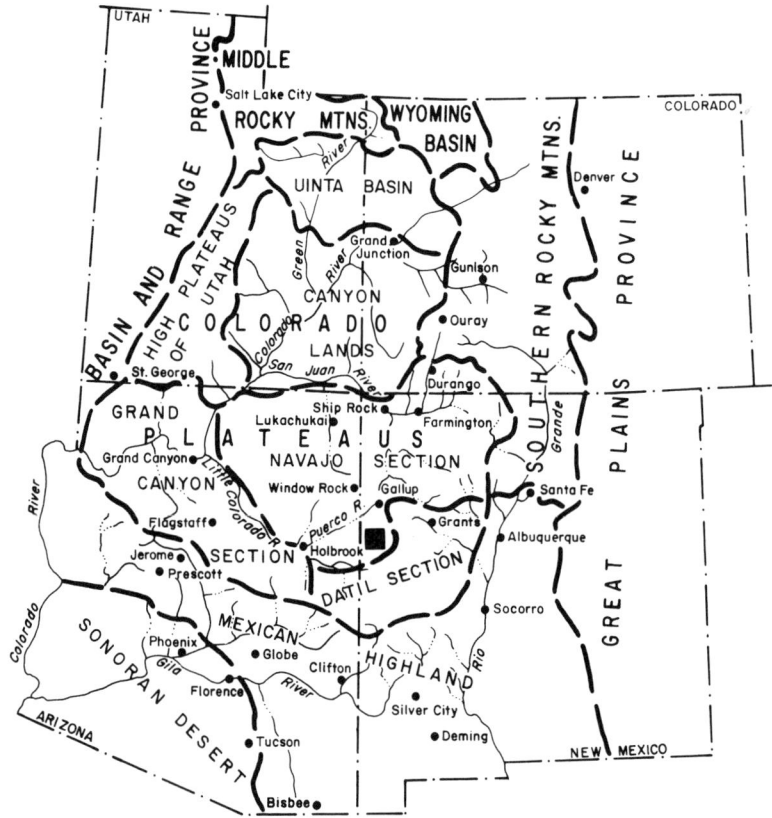

FIG. 22. Location map for the Zuni area of New Mexico.

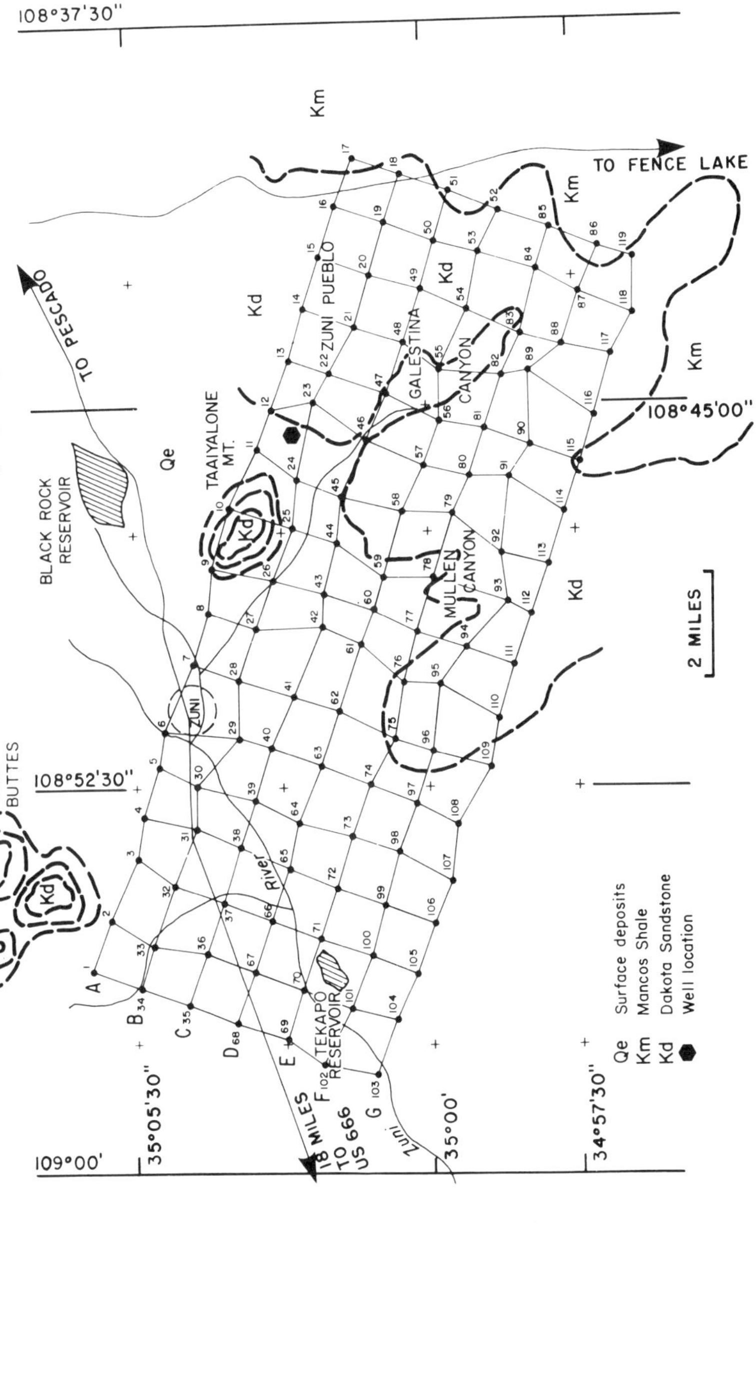

AUDIOFREQUENCY MAGNETOTELLURIC (AMT) SOUNDING 135

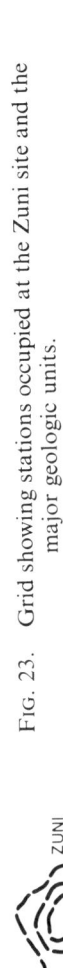

Fig. 23. Grid showing stations occupied at the Zuni site and the major geologic units.

were occupied on a grid with stations at spacings of approximately one mile. The grid of stations occupied is shown in Fig. 23, together with a general geologic map. The sediments are essentially horizontally layered and the resistant Dakota sandstone forms a series of buttes in the area which occupy the high ground in the eastern part of the area. At the easternmost edge of the survey area, the Mancos shale, also of Cretaceous age, overlies the Dakota sandstone. The western part of the area is underlain by a series of formations topped by the Chinle formation which is

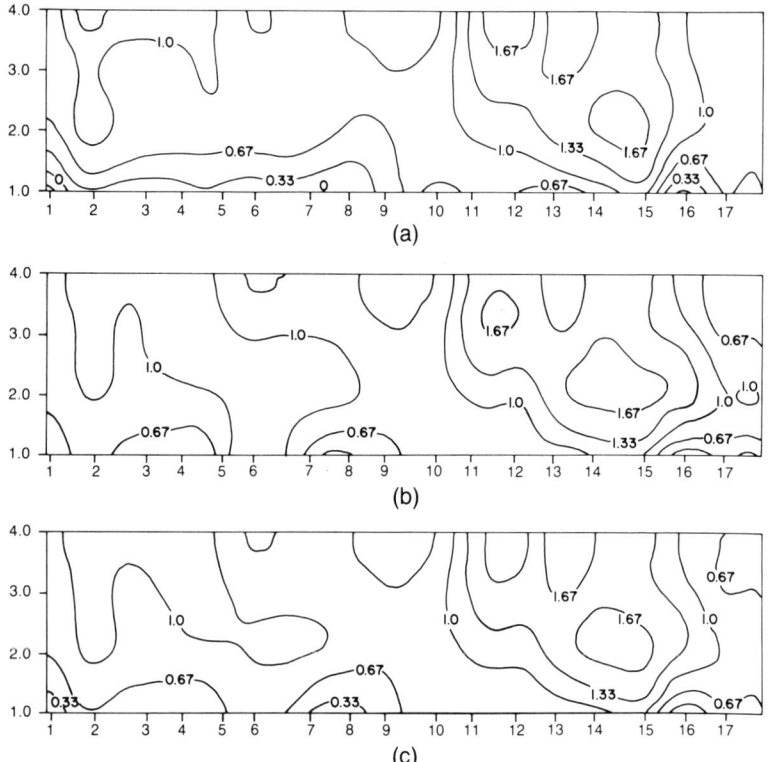

FIG. 24. Pseudosections from line A (shown in Fig. 23). Vertical axis is log frequency. Stations are labelled from 1 to 17. (a) Electrodes oriented N20E; (b) electrodes oriented N70W; (c) average of (a) and (b) determined by squaring the average of the square roots of ρ_a. Contour interval is log ρ_a. The west part of the line is lower resistivity corresponding to the Chinle formation. There are higher resistivities in the eastern part over the Dakota sandstone and then a drop over the Mancos shale at the east end of the line. The uniformity both from station to station and in the two orientations suggests that the layering is essentially horizontal.

composed of shales and sandstones. A drill hole in the area penetrated Precambrian basement at a depth of 767 m.

A series of pseudosections along line A are shown in Fig. 24. In this figure there is essentially no difference between the two electrode orientations, N70W and N20E, which confirms that the area can be considered as a layered case with very little lateral variations. The dominant feature is that the high-frequency resistivities are around 10 Ω m and they decrease with frequency. In the eastern portion of the line, the resistivity rises to about 50 Ω m over the Dakota sandstone and then drops at the easternmost edge to lower values over the Mancos shale. The layering is quite clear as well as the variations associated with the known geology.

These results have also been presented as contour maps of the apparent resistivity at a particular frequency. A high-frequency map and a low-frequency map are shown in Fig. 25. It can be shown that at 8 kHz the

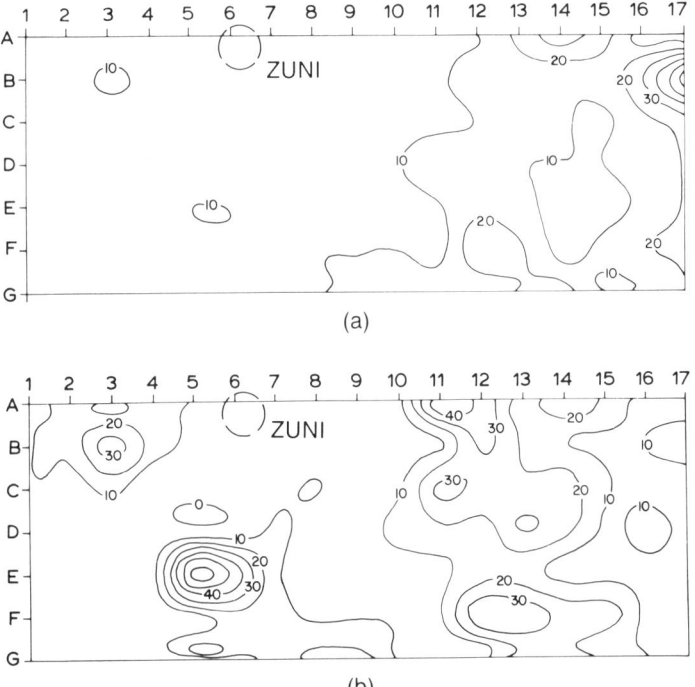

FIG. 25. Contour map (interval 10 Ω m) of the apparent resistivity measured over the grid shown in Fig. 23; lines numbered as in Fig. 23. (a) Frequency of 36 Hz; (b) frequency of 8680 Hz. Note that the surface geology corresponding to the Dakota sandstone is best demonstrated at high frequencies.

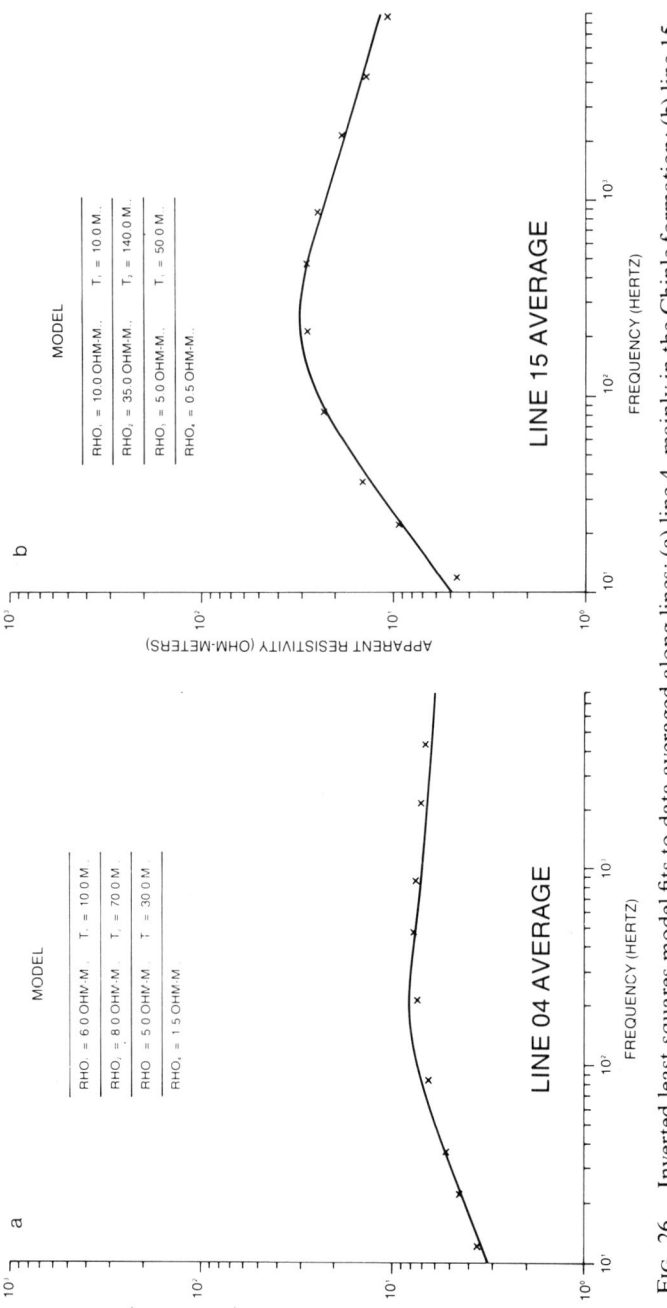

FIG. 26. Inverted least-squares model fits to data averaged along lines: (a) line 4, mainly in the Chinle formation; (b) line 15, mainly in the Dakota sandstone.

higher resistivities are found over the massive Dakota sandstone, and lower resistivities over the Mancos shale and the Chinle formation. At low frequencies, the values continue to be low but there is, as expected, very little correlation with the surface geology.

Some of these data have been inverted. An average along line 4 in the Chinle formation gives the results shown in Fig. 26. The resistivities are very low and drop to $1 \cdot 5 \, \Omega$ m at depth. The net result is that depth penetration is at most 100–200 m even at the lowest frequencies. The corresponding

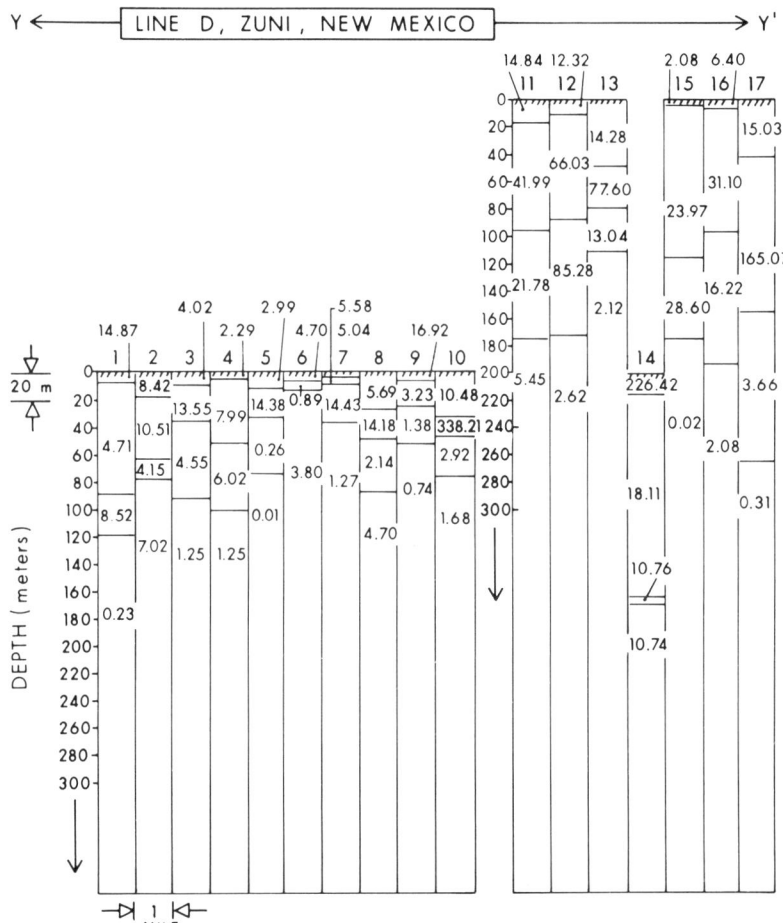

FIG. 27. Results of inverting the data along lines 1–17 (see Fig. 23) to give four layer interpretations at one mile intervals.

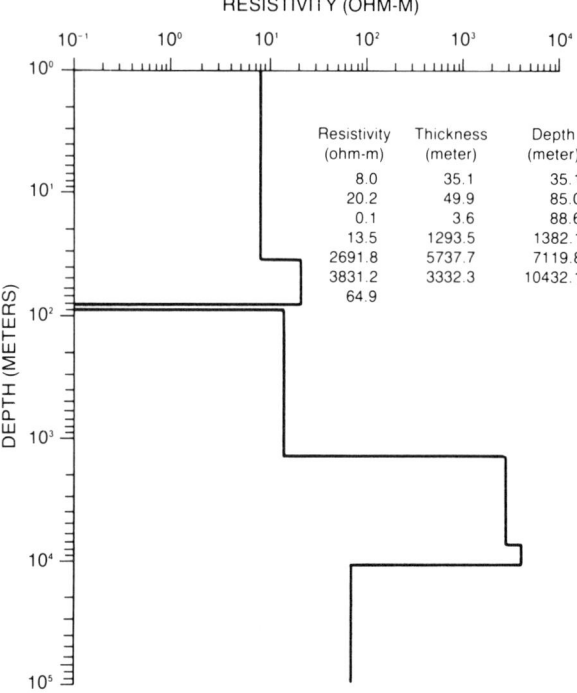

FIG. 28(a). Model derived from fitting data shown in Fig. 28(b). This model is based on a seven-layer case and uses frequencies from 5×10^{-4} Hz to 10^4 Hz. These data are from station 25 (Fig. 23).

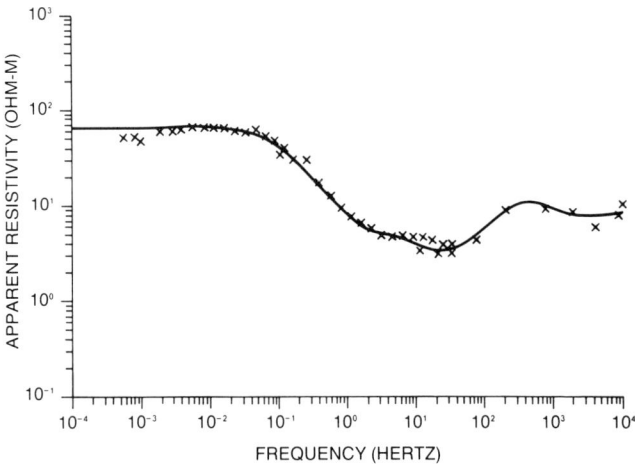

FIG. 28(b). Observations ($\times$) and model fit corresponding to the model of Fig. 28(a).

results over line 15 in the Dakota sandstone show that there is about 150 m of this more resistive unit present. This corresponds well with the known geology. Figure 27 shows the inverted resistivity sections using four-layer inversions for lines 1 to 17.

In the same area a number of stations were occupied for low-frequency MT sounding. This work was done by Woodward Clyde Associates and has been reported by Ander.[33] There was overlap in the frequency range of 12–50 Hz and the agreement in the apparent resistivities was excellent. By combining these results with the AMT data, it was possible to interpret results using eight decades of frequency. Data from station 25 were combined using the two sets of data and a seven-layer inversion model used to invert the results. This has led to the interpreted model of Fig. 28.

Combining these results leads to the model shown in Fig. 28(a). The resistive Precambrian basement is at a depth of about 1400 m. Since this is greater than the drilled depth, it is possible that the upper few hundred metres of the basement has conductive fluids in a fractured and weathered layer and only at a depth of 1·4 km do these fractures close up enough to create the typical high resistivities of the Precambrian. At a depth of 10 km or more the resistivity drops sharply again.

These results show that there is uniform layering in a large part of the geologic column and that it is possible to do rapid mapping of the top 100–200 m using the AMT method. It is thus possible to use the high-frequency AMT data to assist in the interpretation of low-frequency MT data.

3.1.5. Permafrost Mapping

An unusual example of the use of the AMT method was reported by Koziar and Strangway[34,35] who used it for sounding the depth of permafrost in a region of the MacKenzie delta. Frozen material has a very high resistivity and the object was to map the base of the permafrost. This of course is only possible where the permafrost has a great deal of water in it so there is a sharp contrast between the frozen and the unfrozen material. Figure 29 shows a sketch of the area studied and the stations located on two lines. Plots of log resistivity versus log frequency are shown in Fig. 30, which shows that in general there is a nearly straight line relationship. The resistivity is high at high frequency and low at low frequency, giving an impression of a simple two-layer case with a resistive layer over a conductive one, and with a very large contrast such as is seen in the two-layer curves of Fig. 3.

When the curve is a simple 45° line, it is possible to use each data point to determine the key parameters with the results shown in Figs 31 and 32. The

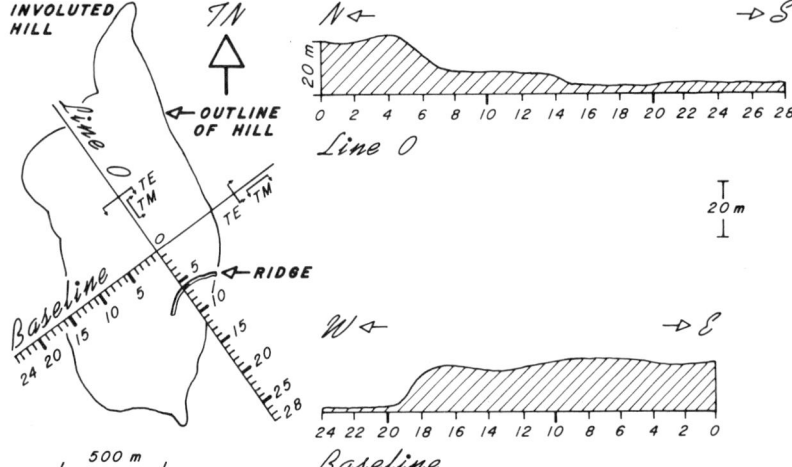

FIG. 29. Location of stations at the Involuted Hill area near Tuktoyaktuk in the McKenzie delta. Two profiles were measured: baseline and line 0. The topography along these is illustrated.

resistivity of the lower layer varies between 10 and 100 Ω m. The resistivity of the permafrost cannot be determined, but is at least 10^3 to 10^4 times as much. The thickness of the permafrost is also determined and gives an answer almost exactly the same as that determined by shallow drilling. In this case then, the AMT method is a simple, rapid coverage technique that has been very effective at determining the permafrost thickness.

3.2. Two-Dimensional Cases

It is very common to encounter large lateral variations in resistivity in the crust. These may arise from many causes such as intrusives, water-filled fracture zones in a resistive region, the presence of massive sulphides or of graphitic sediments in a Precambrian crust. There are many examples of two-dimensional features. Here we shall only discuss two examples: major faulting and massive sulphides. A previous paper[2] described a case history in Montana where it was possible to outline very sharply a massive sulphide zone beneath resistive limestone cover.

3.2.1. Chalk River

An AMT survey was carried out at the Chalk River site (Atomic Energy of Canada Ltd) in connection with the nuclear waste disposal programme.[36,37] Measurements of the scalar apparent resistivity were made at 14 different

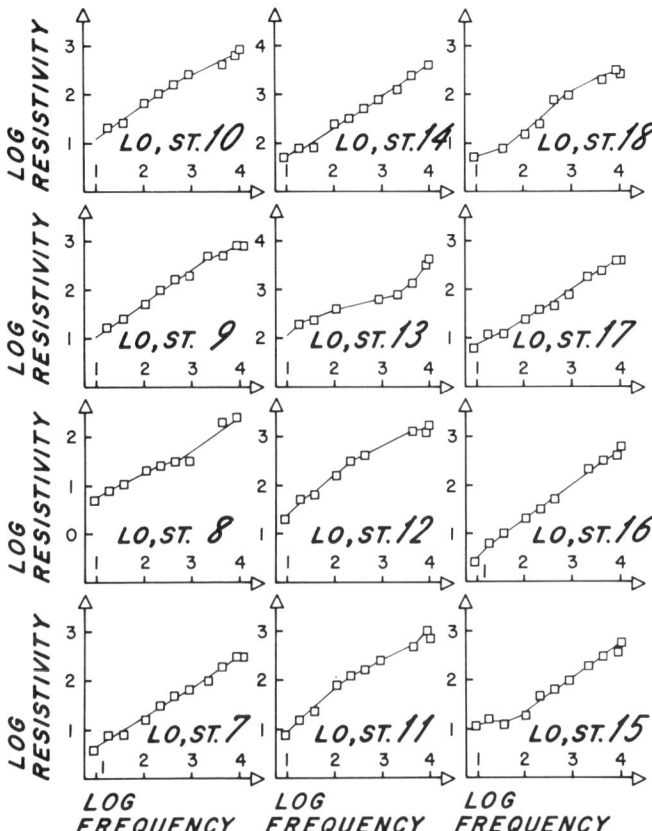

FIG. 30. AMT soundings at 12 stations along line 0 (Fig. 29). Note that in the $\log \rho_a - \log f$ plots these are very nearly straight lines with a slope of 45°. This corresponds to a resistive layer (permafrost) over a conductive layer (fluid rich sediments).

frequencies (12, 22, 36, 83, 213, 428, 858, 2140, 4280, 8570, 10 200, 13 600, 17 800, 18 600 Hz) at each station. The lower 11 frequencies measured at two orthogonal orientations used sources from worldwide thunderstorm activity. The highest three used the US Navy and Omega VLF transmitter stations with the electric sensors in an E–W direction. A second survey was conducted in 1978. The contamination of the natural signals by the nearby power lines was found to be serious. The measurements were made at 60 Hz and at three of its strongest harmonics (180, 300 and 660 Hz) using the power line source instead of natural sources. The electric field was

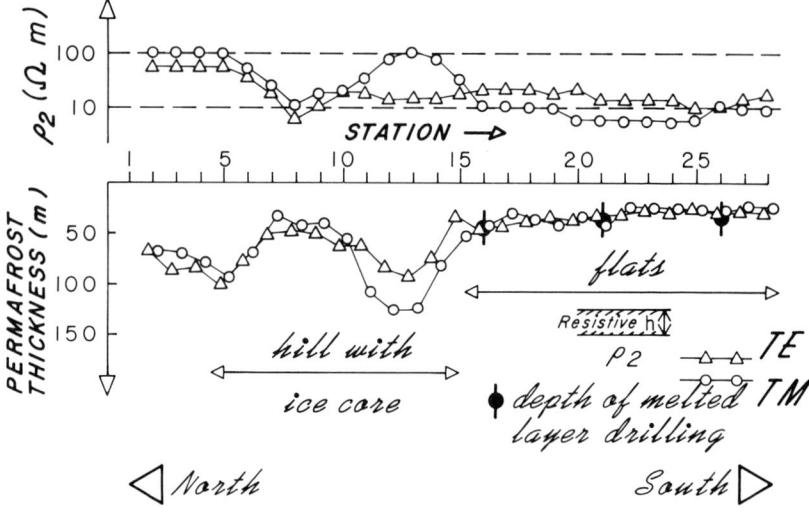

FIG. 31. Profile showing results of fitting a two-layer model at each station along line 0. Open circles and triangles give depths and bottom layer resistivities from orthogonal data sets. Solid circles represent permafrost determined by drilling. Data are in good agreement except where there is a significant lateral variation.

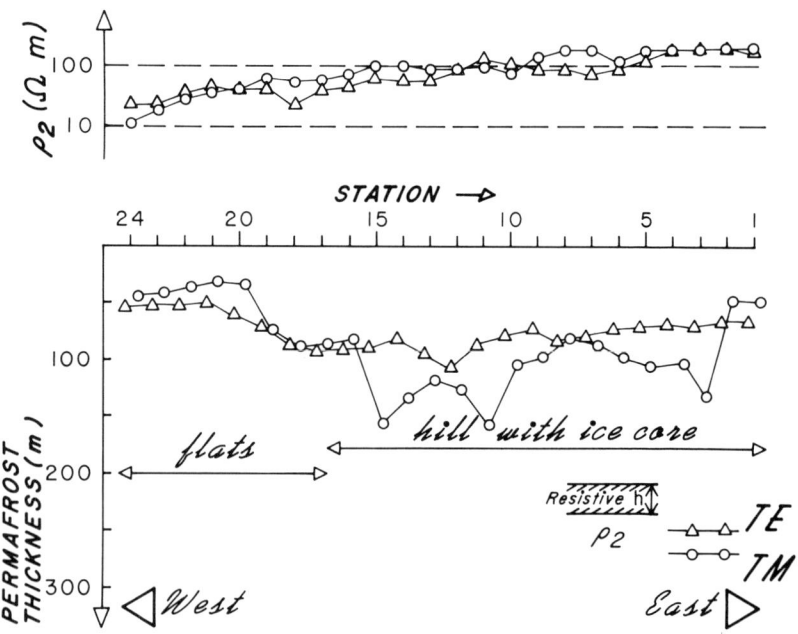

FIG. 32. Profile along baseline showing the interpreted depths and lower-layer resistivities (as in Fig. 31).

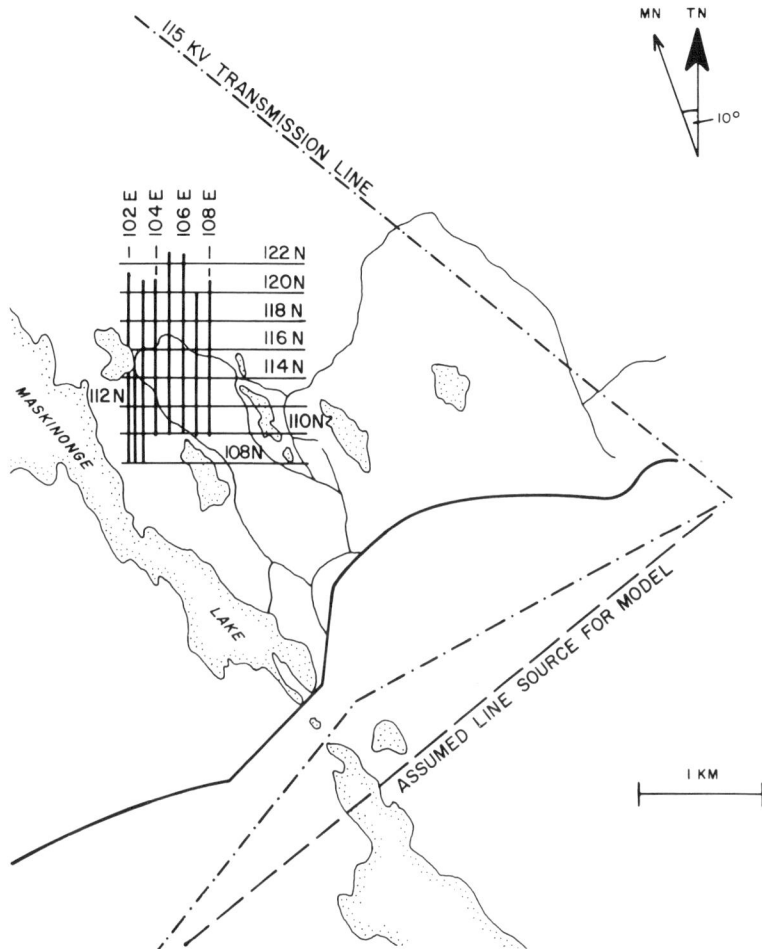

FIG. 33. Location map showing profiles run in a region near Chalk river, Ontario. Power lines used as a source for some of the studies are shown.

measured with a 50 m dipole oriented along the N–S direction and the induction coil was aligned perpendicular to it. This direction for the coil was chosen because it was close to the direction of the major axis of the polarisation ellipse for the horizontal magnetic field at 60 Hz. Stations were spaced at 50 m intervals between 110N and 120N along lines from 102E to 108E forming a grid. The grid, some major geologic features in the area and the location of power lines are shown in Fig. 33. These results are reported in Strangway et al.[25]

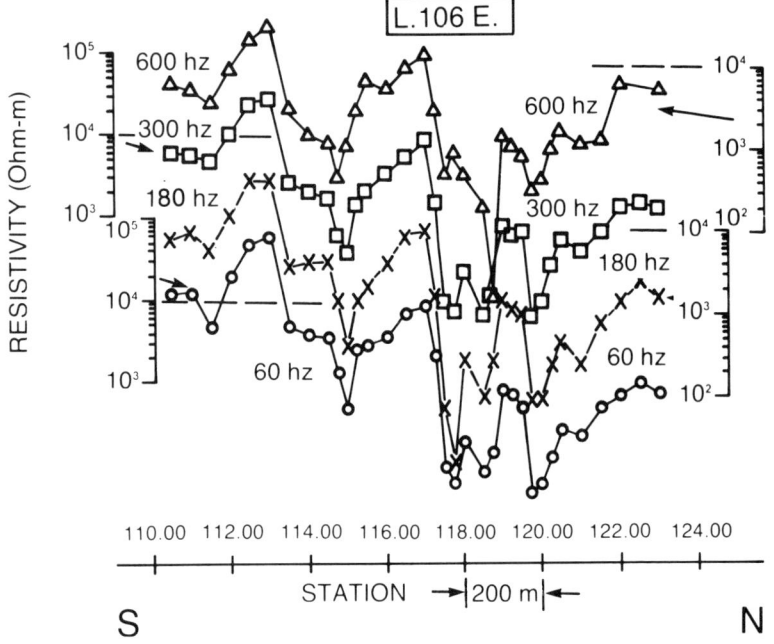

FIG. 34. Profiles of apparent resistivity measured along line 106E (see Fig. 33) at 60 Hz, 180 Hz, 300 Hz and 600 Hz. Note separate resistivity scales used for each profile.

The Chalk River research area is located approximately 200 km west of Ottawa along the Ottawa River, with hilly terrain which is a typical feature of the upper Ottawa valley, there is a relief of the order of 100 m. The region is underlain by complexly-folded crystalline rocks of the Grenville province of the Canadian shield and covered by sand and gravel deposits. Dence and Scott[38] show that the bedrock geology in this area has been proved to be extremely complex by surface sampling and drilling. There is a considerable variation in lithology and degree of fracturing over short distances. The lithologic heterogeneity and structural complexity are typical characteristics of the Grenville province. The main rock unit is a folded sheet of garnetiferous quartz monzonite. It is overlain and underlain by paragneiss and discontinuous pods of metagabbro. There are numerous cross-cutting faults and fractures in the area.

Figure 34 shows the results of the apparent resistivity measurements at four frequencies along line 106 (Fig. 33) and it can be seen that the resistivity is generally high (about $\geq 10\,000\,\Omega\,\text{m}$). This is typical of

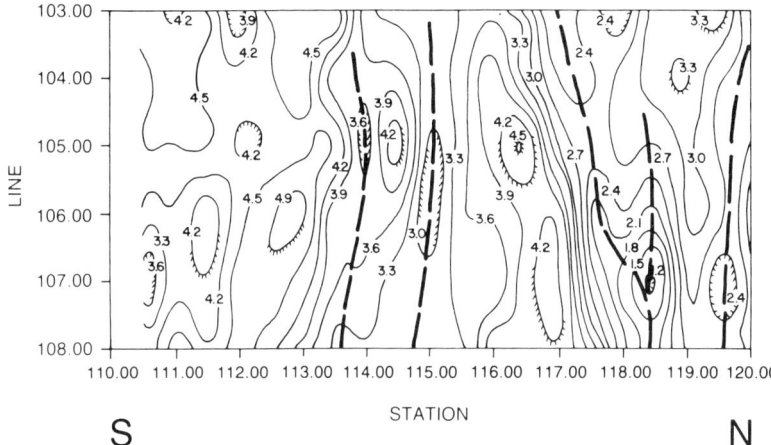

FIG. 35. Contour map of the apparent resistivity measured at 60 Hz over the grid area shown in Fig. 33. Contour interval is one-third $\log \rho_a$. Dashed lines give the location of fault zones.

Precambrian terrain, but sharply defined low resistivities are also found as for example at stations 118 and 120, which show very low resistivities. These features are the locus of major faults as described by Dence and Scott.[38] In Fig. 35 we show a contour map of the apparent resistivity as measured over a region 500 m × 1100 m at a 50 m grid. The faults can be very precisely located by the well-defined resistivity lows. Ideally in two-dimensional studies it would be desirable to have tensor data. However with the rapid, simple, cheap coverage system afforded by scalar AMT, it is possible to do a mapping study and locate boundaries with considerable precision.

The water-filled fractures, which have a very low resistivity in a very resistive medium, are very well-defined. The location of these faults is the same as those located by other electromagnetic means.[36] Even with the simple scalar system it is possible to do an excellent job of locating conductors.

A number of authors have published papers which describe various numerical methods for calculating theoretical profiles over such features.[24,39-42] In a general way, anomalies associated with faults are quite sharp and local and are especially sensitive to the position of the electrodes. The sensitivity is greatest when the electric field is measured parallel to the body, if the depth to the fault or conductor is very shallow (Fig. 36). At greater depths the situation reverses and the sensitivity becomes greatest

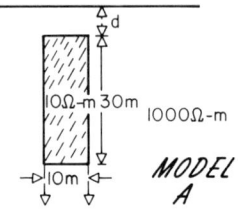

FIG. 36(a). Model used to calculate the results shown in Figs 36(b) and (c).

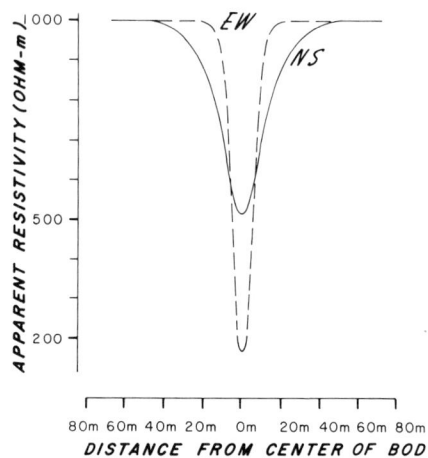

FIG. 36(b). Apparent resistivity calculated for electric arrays normal (N–S) and parallel (E–W) to the body shown in (a). Calculation is based on a frequency of 1000 Hz.

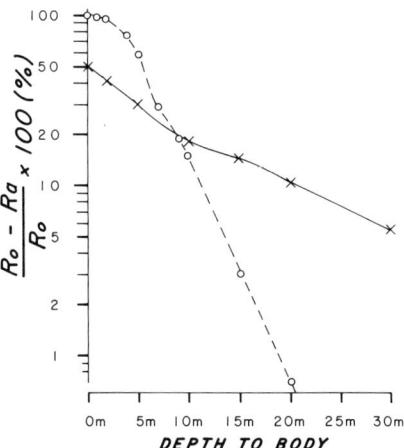

FIG. 36(c). Change in curve shape (peak value R_a, background value R_0) for increasing depth to the body shown in (a). ×, Normal; ○, parallel.

when the electric field is measured normal to the body. Mapping of near-surface faults is therefore best done with measurements parallel to the expected strike direction, while exploring for conductive zones at depth is best done with measurements normal to the expected strike direction.

3.2.2. Cavendish

A second type of two-dimensional feature that is of considerable interest is the exploration for highly conductive massive sulphides. In an earlier paper[2] we described a case in which a massive sulphide zone was clearly delineated at a depth of several hundred metres beneath resistive cover. Massive sulphide zones are typically highly conductive (often with a resistivity of $\leq 1\,\Omega\text{m}$) and are therefore particularly appropriate electromagnetic targets. A target in southern Ontario has been a particularly well-studied deposit, which although non-commercial, has been used for tests of many geophysical methods and has been drilled. AMT data on this deposit have been reported by Strangway and Koziar[4,5] and in this case the method was quite successful at detecting and mapping the boundaries of the deposit. The deposit has been described by Williams et al.[43] and a detailed conventional electromagnetic survey is described by Ward et al.[44]

The location of the deposit and a geological sketch map are shown in Fig. 37. There are two sulphide zones, zone A and zone B, each of which can be detected. The pseudosection shown in Fig. 38 is taken along line C and the two sulphide zones are clearly and sharply resolved. In the parallel profile (E–W) zone B is better defined and in the perpendicular profile zone A is better defined. This is as expected since zone B is at a depth of a few tens of metres of overburden.

An alternative way of showing these same results is given in Fig. 39. This is a set of profiles, frequency by frequency, in which the mean resistivity along each line has been used to normalise the results. The vertical axis is the log of the resistivity at each station divided by the average resistivity. This presentation also clearly shows the two main conductors on line C.

It appears that the AMT approach is useful as a tool for high resolution mapping of massive sulphide zones, as well as for determining the depth to the top of the body.

3.3. Controlled Source Audiofrequency Magnetotellurics (CSAMT)

Audiofrequency magnetotellurics is also now being carried out using artificial sources. Because the detector systems used for natural electric and magnetic fields in the audiofrequency range are remarkably sensitive, it is

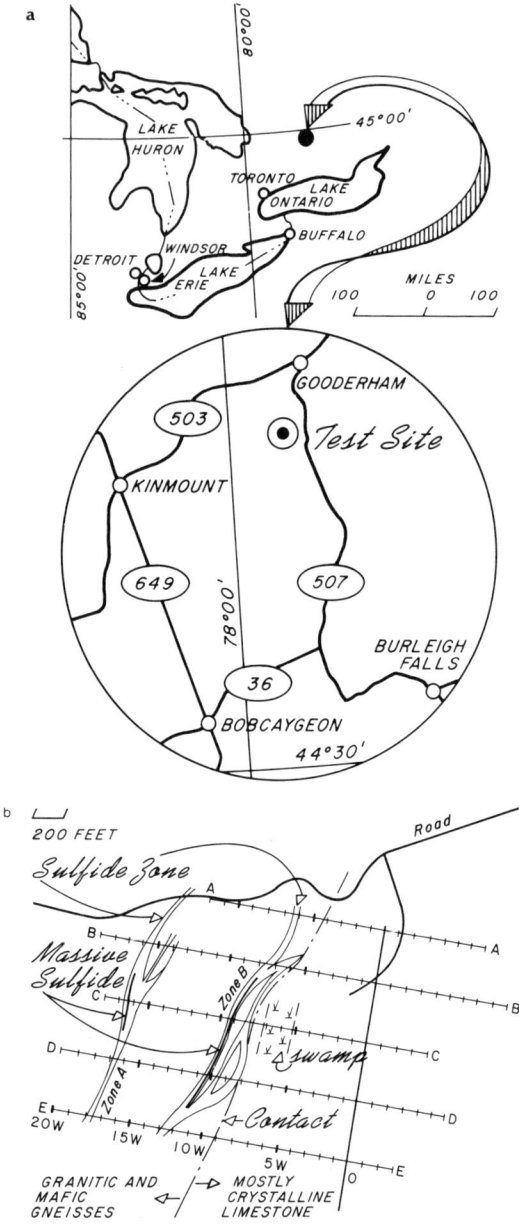

FIG. 37. Location map (a) and geologic sketch map (b) of the Cavendish test site in southern Ontario.

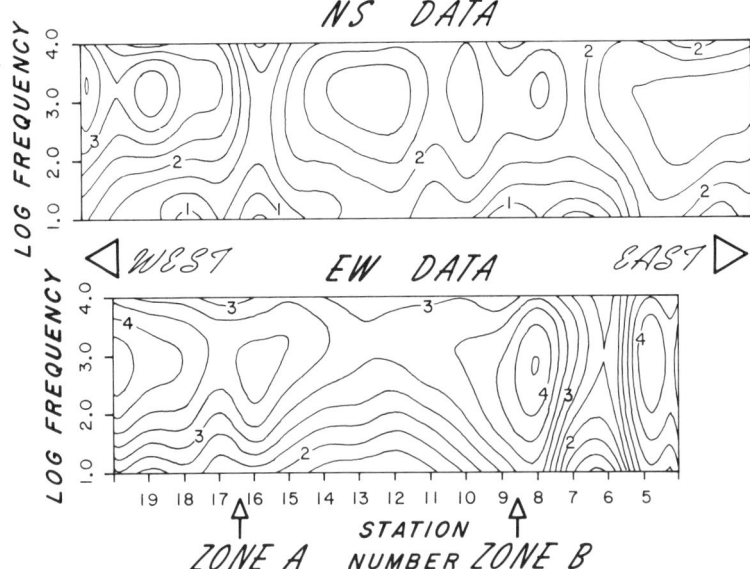

FIG. 38. Pseudosections along line C (Fig. 37). End to end dipole measurements using 100 ft electrode arrays. Contour interval, $0{\cdot}33\log\rho_a$.

possible to use these same analysers at a considerable distance from the source. It is clear that if the detection system is far from the source, then the source can be considered to behave as a plane wave and all the techniques used for AMT analysis and interpretation can be used. The advantage in terms of the quality of the data is considerable although the field logistics become much more cumbersome, costly and slow. The rapid coverage afforded by natural source work is considerably reduced.

Goldstein and Strangway[14] reported on extensive theoretical model and field work for a source using a grounded electric dipole as the power source. This study showed that, provided the source was three or more skin depths from the measurement site, the approximation of a plane wave held to a high degree of accuracy. This means that straightforward one- and two-dimensional interpretation techniques are valid.

The authors calculated the fields (magnetic and electric) expected as a function of distance in a half-space (and in a two-layer earth). A programme of field measurements was then carried out at the Bonneville salt flats in Utah for comparison. The Bonneville salt flats are known to be a very good half-space with a resistivity of $0{\cdot}25\,\Omega\,\text{m}$. Comparison between the

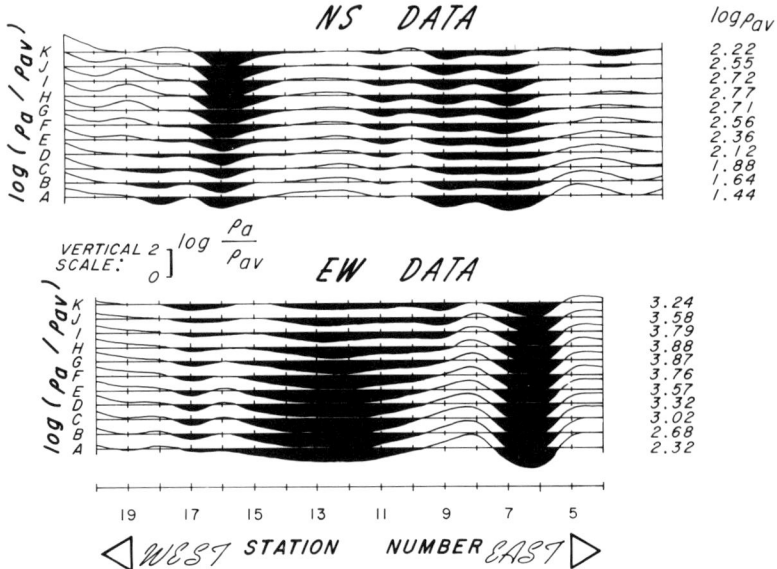

FIG. 39. Residual profiles along line C. Average value at each frequency lettered from high frequency to low frequency (10 000 Hz labelled K and 10 Hz labelled A) is used as the reference (see numbers at right-hand side) and plot made of $\log(\rho_a/\rho_{av})$ at each station along the line and then interpolated to give smooth profile. Shaded regions have below average resistivity.

calculated and measured field strength is given in Fig. 40. This shows that the agreement is excellent. By combining the measured values of E_x, H_y and E_y, H_x it is possible to calculate the apparent resistivity on the assumption of a plane wave source. This has been done for the salt flats and the results are shown in Fig. 41. The regions shown by shading give results that are very close to $0.25\ \Omega\,m$ provided one is more than three skin depths from the source. This simple set of measurements carried out at 100 Hz using a 100 Hz square wave source shows that the CSAMT approach can give quite accurate resistivity measurements.

Other examples are reported in the paper by Goldstein and Strangway[14] including a report on a study of the Cavendish massive sulphide deposit. Van Blaricom reported at the Edinburgh meeting on extensive field experience with the method, illustrated by a large number of examples of successful mapping programmes. The source used was a large loop laid on the surface of the ground. Sandberg and Hohmann[15] have recently reported on a case history in geothermal exploration in Utah. In this report

AUDIOFREQUENCY MAGNETOTELLURIC (AMT) SOUNDING

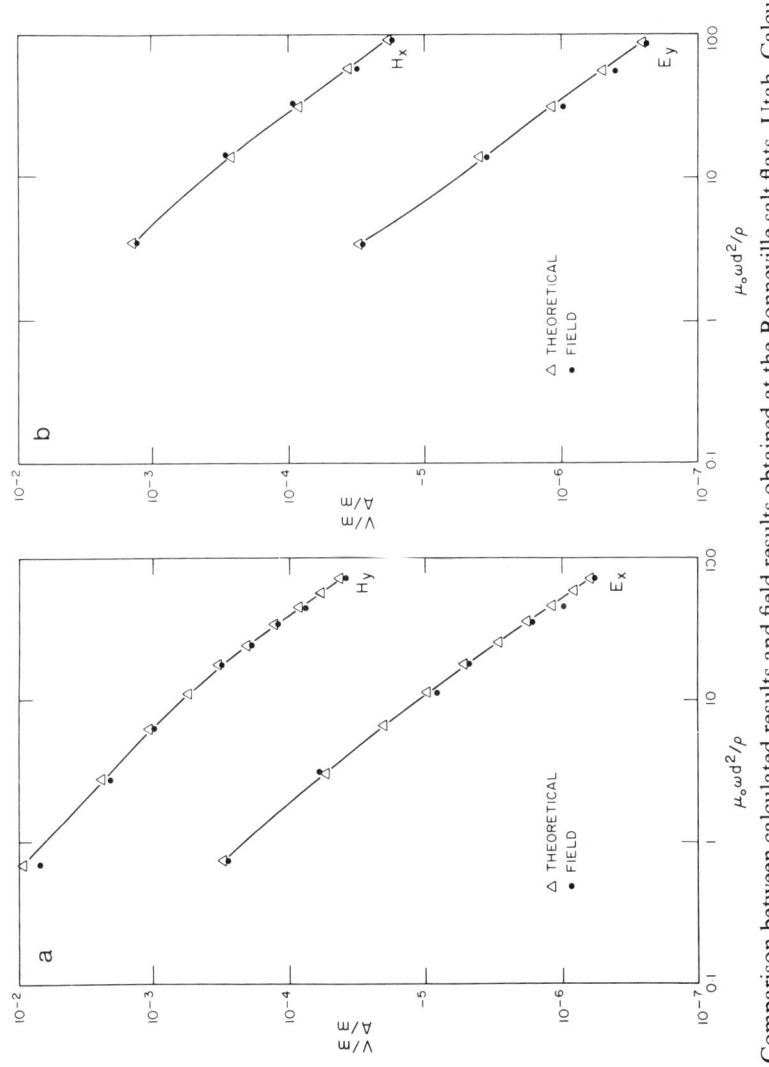

FIG. 40. Comparison between calculated results and field results obtained at the Bonneville salt flats, Utah. Calculations are based on a resistivity of 0·25 Ω m. (a) E_x, H_y along the y axis (normal to the source dipole); (b) E_y, H_x along a line 30° from the dipole axis.

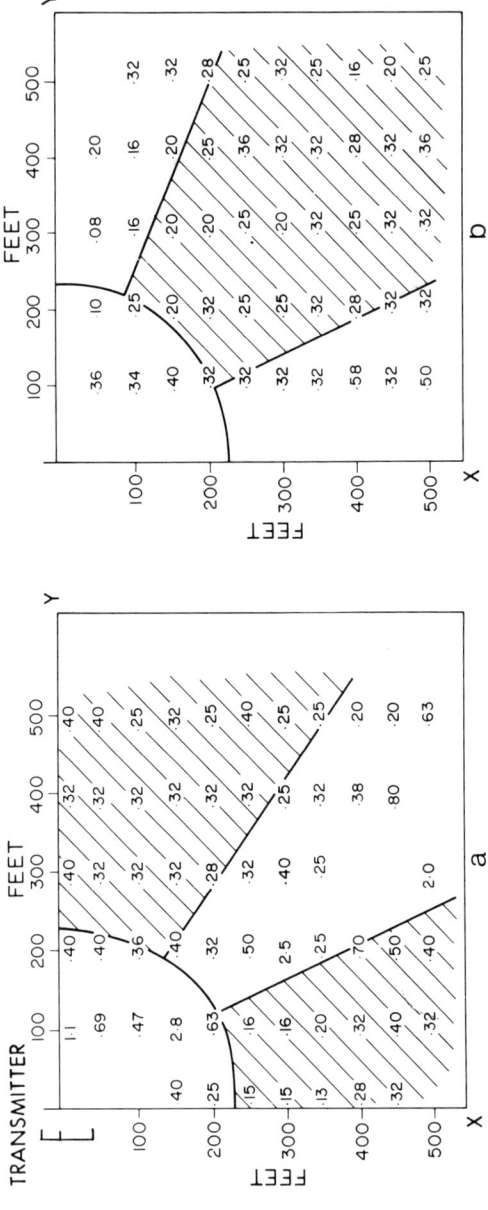

FIG. 41. Resistivities measured in a quadrant around the source dipole at the Bonneville salt flats. The circle marks the three skin depth range. The shaded regions give the best results for the following combinations: (a) E_x/H_y and (b) E_y/H_x (transmitter oriented along the vertical axis).

they used frequencies of 32, 98, 977 and 5208 Hz using a grounded electric dipole source. They developed a series of pseudosections and compared these with conventional dipole-dipole resistivity studies. By carrying out a grid array of observations they were able to utilise three-dimensional modelling techniques to interpret their results. They suggest that the plane wave approximation is valid for distances of more than three skin depths for broadside arrays and of more than five skin depths for collinear arrays. They conclude the CSAMT procedure is effective for high-resolution studies and that it is quite efficient in terms of mapping coverage when compared with conventional resistivity methods.

4. CONCLUSIONS

The audiofrequency magnetotelluric sounding has proven itself to be a useful tool for carrying out high-resolution resistivity mapping. The use of natural sources means that the quality of the data is limited, but it is a fast and simple way to map in areas of high resistivity contrast. In regions where the resistivity varies little laterally, it can produce quite accurate results and the simple scalar method in current usage can lead to accurate inversions. In regions where the resistivity varies laterally, the fact that mapping can be done quite efficiently means that it is possible to outline contacts, faults, sulphide zones, etc., with considerable precision. When tensor systems become available they should improve the quality of data and hence the interpretation of the results, although it will be at considerable cost in terms of the ability for rapid mapping coverage.

Reported applications include shallow mapping in conducting sediments, Precambrian crustal sounding, sulphide exploration, mapping sites for nuclear waste disposal, exploring through high-resistivity cover and geothermal studies. The possible applications are as great as those of any other electrical technique provided high precision data are not needed.

Controlled source AMT, in which a dipole is placed several skin depths from the observation site, yields much more precise data. Experience with this technique is now developing and shows that this approach is also useful.

Plane wave sources and the use of frequencies down to 10 Hz, mean that this method has better depth-penetration capability than conventional electromagnetic exploration methods and that is is a useful supplement to low-frequency deep-sounding techniques.

ACKNOWLEDGEMENTS

The research and field work discussed in this review has been supported by a number of organisations and individuals. I express appreciation to: Kennecott Copper Corporation; the Natural Sciences and Engineering Research Council; the Defence Research Board; Texasgulf; W. G. Wahl Ltd; Energy, Mines and Resources (Geological Survey of Canada, Earth Physics Branch and the Polar Continental Shelf Project); the Department of Supplies and Services; Atomic Energy of Canada Ltd; Woodward Clyde; Los Alamos National Laboratories; the Saskatchewan Mining Development Corporation; GTE Sylvania; the US Navy. All of these organisations have either supported research on the method or supported various field programmes. It is also important to single out a number of individuals who have played an important role—R. C. Holmer, C. Swift, M. Goldstein, A. Koziar, A. Kryzan, A. Gubins, D. Hsu and M. Ilkisik. In particular, J. D. Redman has been instrumental in developing and utilising the system.

REFERENCES

1. CAGNIARD, L. (1953) Basic theory of the magnetotelluric method of geophysical prospecting. *Geophysics* **18**, 605–35.
2. STRANGWAY, D. W., SWIFT, C. M. and HOLMER, R. C. (1973) The application of audio frequency magnetotellurics (AMT) to mineral exploration. *Geophysics* **38**, 1159–75.
3. HOOVER, D. B., FRISCHKNECHT, F. C. and TIPPINS, C. L. (1976) Audio magnetotelluric sounding as a reconnaissance exploration technique in Long Valley, California. *J. Geophys. Res.* **81**, 801–9.
4. HOOVER, D. B., LONG, C. L. and SENTERFIT, R. M. (1978) Some results from audiomagnetotelluric investigations in geothermal areas. *Geophysics* **43**, 1501–14.
5. MABEY, D. R., HOOVER, D. B., O'DONNELL, J. E. and WILSON, C. W. (1978) Reconnaissance geophysical studies in the geothermal system in southern Raft River Valley, Idaho. *Geophysics* **43**, 1470–84.
6. LONG, C. L. and KAUFMAN, H. E. (1980) Reconnaissance geophysics of a known geothermal resource area, Weiser, Idaho and Vale, Oregon. *Geophysics* **45**, 312–22.
7. NGOC, PHAM VAN, BOYER, D. and CHOTEAU, M. (1978) Cartographic des 'pseudo-resistivites apparentes' par profilage tellurique–tellurique associe a la magneto-tellurique. *Geophys. Prosp.* **26**, 218–46.
8. BENDERITTER, Y., HERISSON, C., KORHONEN, H. and PERNU, T. (1978) Magnetotelluric experiments in northern Finland. *Geophys. Prosp.* **26**, 565–71.
9. DUPIS, A. and ILICETO, V. (1974) An example of rapid magnetotelluric

investigation of faulted structures: the Carboli area (Lardarello, Italy). *Boll. di Geofis. Teorica ed Applicata* **16**, 125–36.
10. DUPIS, A., ILICETO, V. and NORINELLI, A. (1974) First magnetotelluric measurements on Lardarello site. *Boll. di Geofis. Teorica ed Applicata* **16**, 137–52.
11. SLANKIS, J. A., TELFORD, W. M. and BECKER, A. (1972) 8-Hz telluric and magnetotelluric prospecting. *Geophysics* **37**, 862–78.
12. GUINEAU, B. (1973) Survey data resulting of 'high frequency' magnetotelluric prospecting (ranging from 10 kHz to 500 kHz) over inhomogeneities and geological structures located under a very thin overburden. *Geophys. Prosp.* **21**, 598 (abstract only).
13. GOLDSTEIN, M. A. (1971) Magnetotelluric experiments employing an artificial dipole source. Unpublished Ph.D. thesis, Department of Physics, University of Toronto.
14. GOLDSTEIN, M. A. and STRANGWAY, D. W. (1975) Audio frequency magnetotellurics with a grounded electric dipole source. *Geophysics* **40**, 669–83.
15. SANDBERG, S. K. and HOHMANN, G. W. (1982) Controlled source audiomagnetotellurics in geothermal exploration. *Geophysics* **47**, 100–16.
16. WAIT, J. R. (1962) Theory of magnetotelluric fields. *J. Res. NBS* **66D**, 509–41.
17. VOZOFF, K. (1972) The magnetotelluric method in the exploration of sedimentary basins. *Geophysics* **37**, 98–141.
18. YUNGUL, S. H. (1961) Magnetotelluric sounding three-layer interpretation curves. *Geophysics* **26**, 465–70.
19. MADDEN, T. R. and SWIFT, JR., C. M. (1969) Magnetotelluric studies of the electrical conductivity structure of the crust and upper mantle. In *The Earth's Crust and Upper Mantle*, AGU Monograph 13, pp. 469–79.
20. VOZOFF, K. and STRANGWAY, D. W. (1972) Magnetotellurics in Exploration. In *Proceedings of a United Nations Conference on 'New Techniques of Mineral Exploration with Emphasis on Geophysical Methods'*, Gordon and Breach, New York.
21. CANTWELL, T. (1960) Detection and analysis of low frequency magnetotelluric signals. Ph.D. thesis, Department of Geology, Massachusetts Institute of Technology.
22. BOSTICK, JR., F. S. and SMITH, H. W. (1962) Investigation of large-scale inhomogeneities in the earth by the magnetotelluric method. *Proc. Inst. Radio Eng.* **50**, 2339–46.
23. SWIFT, C. M. (1967) A magnetotelluric investigation of an electrical conductivity anomaly in the southwestern United States. Ph.D. thesis, Department of Geology and Geophysics, Massachusetts Institute of Technology.
24. SWIFT, C. M. (1971) Theoretical magnetotelluric and Turam response from two-dimensional inhomogeneities. *Geophysics* **36**, 38–52.
25. STRANGWAY, D. W., REDMAN, J. D. and MACKLIN, D. (1980) Shallow electrical sounding in the Precambrian crust of Canada and the United States. In *The Continental Crust and its Mineral Deposits*, Geological Association of Canada, special paper No. 20.

26. ILKISIK, O. M., HSU, D. T., REDMAN, J. D. and STRANGWAY, D. W. (1982) Surface electromagnetic mapping in selected positions of Northern Ontario. Accepted in Miscellaneous Reports of the Ontario Geological Survey.
27. DUNCAN, P. M., HWANG, A., EDWARDS, R. N., BAILEY, R. C. and GARLAND, G. D. (1980) The development and applications of a wide band electromagnetic sounding system using a pseudo-noise source. *Geophysics* **45**, 1276–96.
28. SANDFORD, B. V. (1969) Geology, Toronto–Windsor area, Ontario: Map 1263a. Geological Survey of Canada, Department of Energy, Mines and Resources, Ottawa.
29. HSU, D. (1980) One-dimensional and two-dimensional interpretation of audiomagnetotelluric data. Unpublished M.Sc. thesis, Department of Geology, University of Toronto.
30. GUBINS, A. G. (1979) Magnetic and audiofrequency–magnetotelluric (AMT) investigations of buried astroblemes in the Williston Basin. Unpublished M.Sc. thesis, Department of Geology, University of Toronto.
31. RANKIN, D. and KAO, D. (1978) The delineation of the Superior–Churchill transition zone in the Canadian shield. *CSEG J.* **14**, 50–4.
32. KOZIAR, A. and STRANGWAY, D. W. (1978) Permafrost mapping by audio frequency magnetotellurics. *Can. J. Earth Sci.* **15**, 1539–46.
33. ANDER, M. (1981) Geophysical study of the crust and upper mantle beneath the Central Rio Grande rift and adjacent Great Plains and Colorado plateau. Ph.D. thesis, Department of Geology, University of New Mexico.
34. KOZIAR, A. and STRANGWAY, D. W. (1975) Magnetotelluric sounding of permafrost. *Science* **190**, 566–8.
35. KOZIAR, A. and STRANGWAY, D. W. (1978) Shallow crustal sounding in the Superior province by audio frequency magnetotellurics. *Can. J. Earth Sci.* **15**, 1701–11.
36. STRANGWAY, D. W. (1979) Geophysical methods for selection and *in situ* testing of waste disposal sites. Geol. Surv. Canada, paper 79, p. 10.
37. STRANGWAY, D. W. (1980) Geophysical methods and toxic waste disposal. *Geoscience Canada* **7**, 30–2.
38. DENCE, M. R. and SCOTT, W. J. (1979) The use of geophysics in the Canadian radioactive waste disposal program, with examples from the Chalk River research area. *Geoscience Canada* 190–3.
39. D'ERCEVILLE, I. and KUNETZ, G. (1962) The effect of a fault on the earth's natural electromagnetic field. *Geophysics* **27**, 651–65.
40. RANKIN, D. (1961) The magnetotelluric effect on a dike. *Geophysics* **28**(5), 666–76.
41. JONES, D. L. and PASCOE, L. J. (1971) A general computer program to determine the perturbation of alternating electric currents in a two-dimensional model of a region of uniform conductivity with an imbedded inhomogeneity. *Geophys. J. Roy. Astr. Soc.* **23**, 3–30.
42. SILVESTER, P. and HASLAM, C. R. S. (1972) Magnetotelluric modelling by the finite element method. *Geophys. Prosp.* **20**, 872–91.
43. WILLIAMS, D. A., SCOTT, W. J. and DYCK, A. V. (1975) Cavendish Township geophysical test range, 1975 diamond drilling. Geological Survey of Canada, paper 74, p. 62.

44. WARD, S. H., PRIDMORE, D. F., RIJO, L. and GLENN, W. E. (1974) Multispectral electromagnetic exploration for sulfides. *Geophysics* **39**, 666–82.
45. STRANGWAY, D. W. and KOZIAR, A. (1979). Audiofrequency magnetotelluric sounding: a case history at Cavendish geophysical test range. *Geophysics* **44**, 1429–46.
46. BUDDEN, K. G. (1961) *The Wave Guide Mode Theory of Wave Propagation*, Prentice Hall Inc., New York.

Chapter 5

DETECTION AND MAPPING OF TUNNELS AND CAVES

T. E. OWEN

*Southwest Research Institute,
San Antonio, Texas, USA*

SUMMARY

Several geophysical exploration methods have been applied to the problem of detecting and mapping underground cavities with practical success. The most productive techniques are those for which the cavity target exhibits the greatest physical contrast with respect to the measurable parameters of the host geologic medium. Development of these methods has required a fresh insight into the geophysical concepts being applied, and further optimisation of results has required an upgrading of the sensitivity, resolution and data acquisition accuracy of the methods in contrast with their conventional capabilities. In cases where the cavity is relatively small compared with its depth, surface survey methods provide a search-mode capability whereas borehole methods serve in a complementary way to check suspected target anomalies. The geophysical techniques most widely used include electrical resistivity, seismic, electromagnetic and gravity surveys applied at the surface and in drill holes. The performance capabilities of these methods are discussed in regard to certain specific cave and tunnel targets with the aim of familiarising the reader with their potential adaptation to similar applications.

1. CAVITY DETECTION APPLICATIONS

A variety of engineering and scientific applications can benefit from geophysical field surveys capable of detecting and mapping tunnels and caves. For example, both natural and man-made underground cavities may cause ground subsidence problems that affect existing cultural structures on the surface. Likewise, major new construction projects such as nuclear power plants require stringent foundation-evaluation surveys to ensure that subsidence problems will not develop. The integrity of water-impoundment dams is seriously threatened by underlying cavities which may promote underground erosion causing instabilities in the dam structure. Flooded abandoned mine workings can pose inundation hazards to other nearby underground mining operations. Water run-off channelled through volcanic lava tubes can cause surreptitious flooding under extreme conditions. Subversive intrusion tunnelling can impose military threats to political stability.

Geophysical methods of tunnel and cavity detection also apply to hydrogeological surveys for water sources, surveys for archaeological chamber tombs and in speleologic exploration. Other applications of these techniques also exist in probing and evaluating *in situ* rubble zones associated with oil shale conversion and in defining the size and shape of solution cavities in salt domes.

The underground cavity targets of interest consist of natural and man-made anomalies typically within a depth of about 100 m of the ground surface. The most abundant natural targets are limestone solution cavities and soil sinkholes in karstic terrain; lava tubes are also natural underground cavity targets of interest. More prevalent, however, are various man-made cavities such as underground mine workings and near-surface entry shafts, intrusion tunnels of tactical military or defensive origin, and archaeological chamber tombs. Deeper and larger cavity targets include rubble zones associated with *in situ* oil shale processing, and solution cavities in salt domes as they relate to brine-recovery operations and to underground petroleum storage.

Reconnaissance surveys to detect underground cavities usually involve surface geophysical techniques applied over bounded geographical areas where search and detection is of concern. Search strategies are typically based upon line traverses or survey grid networks laid out to cover the search area of interest. Importantly, however, test borings and other drilling efforts are employed as directed by the interpreted survey results to physically verify the suspected cavity anomalies. Because of high drilling costs, geophysical survey methods have been developed to reduce the overall amount of drilling and to improve the placement of the test borings for more efficient target verification and mapping.

DETECTION AND MAPPING OF TUNNELS AND CAVES   163

In addition to surface survey techniques, if a drilling programme is in progress, several borehole geophysical probing methods are available for examining the underground zones surrounding or lying between the drill holes. Similarly, in underground mining applications certain specialised techniques are available for operation in horizontal drill holes or in the accessible underground areas of the mine for probing into the surrounding geological formations.

The relatively high physical contrast between an air-filled void and the surrounding geological material makes these cavity targets susceptible to detection by several indirect sensing methods. Water-filled cavities also offer a high contrast with surrounding geological materials and generally produce a different geophysical response from that of an air-filled cavity so that one can discriminate between empty and water-flooded cavities.

The various geophysical techniques applicable to detection and mapping of cavities and tunnels have evolved as a result of a particular problem or need. However, many of these methods are applicable to more than one cavity detection requirement. Therefore, in the discussions to follow the particular applications which originated the methods will be emphasised and their potential applications in other cavity detection problems will also be mentioned. These discussions will emphasise the detection and mapping of caves and tunnels which are free of any particular man-made contents such as ferromagnetic rails, wires, or activated power cables which might otherwise be used as a basis for indirectly detecting the cavity.

2. GEOPHYSICAL TECHNIQUES

Various geophysical exploration techniques have been evaluated and applied to the problem of detecting and locating tunnels and cavities. The most widely used surface methods include electrical resistivity, gravity and seismic techniques. More recently, however, controlled-source audio-magnetotelluric and ground-penetrating VHF radar techniques have been developed to meet certain cavity-detection requirements. The most productive borehole methods of cavity detection include hole-to-hole and hole-to-surface seismic transmission, hole-to-hole electromagnetic transmission, borehole VHF radar, hole-to-hole resistivity, and most recently, borehole gravimetry. Several other techniques employing magnetic, thermal and acoustic measurements have also been tried in various tunnel and cave detection surveys with relatively limited success.

3. ELECTRICAL RESISTIVITY METHODS

Earth resistivity measurements have been the most extensively applied approach to tunnel and cave detection problems and have met with various degrees of success. An early application of this method is described by Palmer[1-4] in reference to the location of subterranean caves. This method employed a symmetrical four-electrode technique in which the half-array electrode spacing ratio was held constant as the array was expanded to provide depth soundings. The electrode geometry adopted in this case provides a basis for deriving the depth of the cavity from a single resistivity sounding centred over the target. The work of Palmer is significant in that he recognised the relatively unique problem associated with the detection of localised resistive anomalies and devised appropriate methodology adapted to the need. In a similar practical approach to this problem, Bristow[5] adopted the pole–dipole electrode array in a manner which allowed direct graphical interpretation of limestone solution cave targets in approximate depth, position and size.

An extension of this survey technique was successfully demonstrated by Bates[6] to detect solution cavities in karst terrain with the potential application of detecting solution erosion problems under water reservoir dams. Further strides in the development and application of pole–dipole methods have been accomplished by the Southwest Research Institute to detect soil sinkhole cavities affecting ground stability along highways,[7] to detect and map subversive underground tunnels,[8-10] and to locate abandoned coal mine workings.[11]

The general four-electrode method of earth resistivity measurement can be specialised to represent the various geoelectrical survey geometries used in practice. Consider the electrode layout shown in Fig. 1 in which the current injection electrodes $C_1(x, y)$ and $C_2(x, y)$, separated by a distance $2a$, and the potential measurement electrodes $P_1(x,y)$ and $P_2(x,y)$, separated by a distance $2b$, are oriented arbitrarily on a flat halfspace having uniform resistivity, ρ.

For a homogeneous halfspace, the resistivity of the medium is expressed by

$$\rho = \frac{2\pi}{\dfrac{1}{r_1} - \dfrac{1}{r_2} - \dfrac{1}{r_3} + \dfrac{1}{r_4}} \left(\frac{V}{I}\right)$$

$$= K\left(\frac{V}{I}\right) \qquad (1)$$

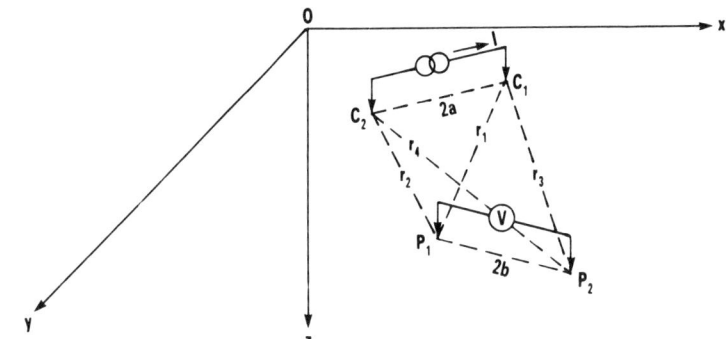

FIG. 1. Geometry of four-electrode array for earth resistivity measurements.

where I = injected current at electrode C_1 and returned to electrode C_2;
V = potential at electrode P_1 with respect to electrode P_2;

$$K = \frac{2\pi}{\dfrac{1}{r_1} - \dfrac{1}{r_2} - \dfrac{1}{r_3} + \dfrac{1}{r_4}}$$

= a geometric factor by which the measured ohmic factor (V/I) for any arbitrary electrode layout is converted to resistivity of the medium. Distances r_1, r_2, r_3 and r_4 are defined in Fig. 1.

For the usual case where the halfspace is not composed of a uniform resistivity material, the resistivity derived using eqn (1) is termed the apparent resistivity,

$$\rho_a = K\left(\frac{V}{I}\right) \qquad (2)$$

in which perturbations may occur in the measured ohmic factor (V/I) as a result of subsurface inhomogeneities. Therefore, the problem of practical resistivity data analysis and interpretation is one of inferring the electrical contrasts representing subsurface structure from spatial distributions of apparent resistivity observed on the surface.

Table 1 summarises several resistivity arrays which have been used in practice, showing their specialised geometrical layouts and geometrical factors for deriving apparent resistivity. The in-line array configurations are the most widely used because of the ease with which the field grid can be laid out and surveyed. The Wenner array has been one of the most commonly used arrays because of its simple geometry and apparent resistivity computation. However, it is less sensitive to localised anomalies than the Schlumberger, central electrode, pole–bipole, or bipole–bipole

TABLE 1
SUMMARY OF EARTH RESISTIVITY ELECTRODE ARRAYS AND GEOMETRIC FACTORS FOR APPARENT RESISTIVITY

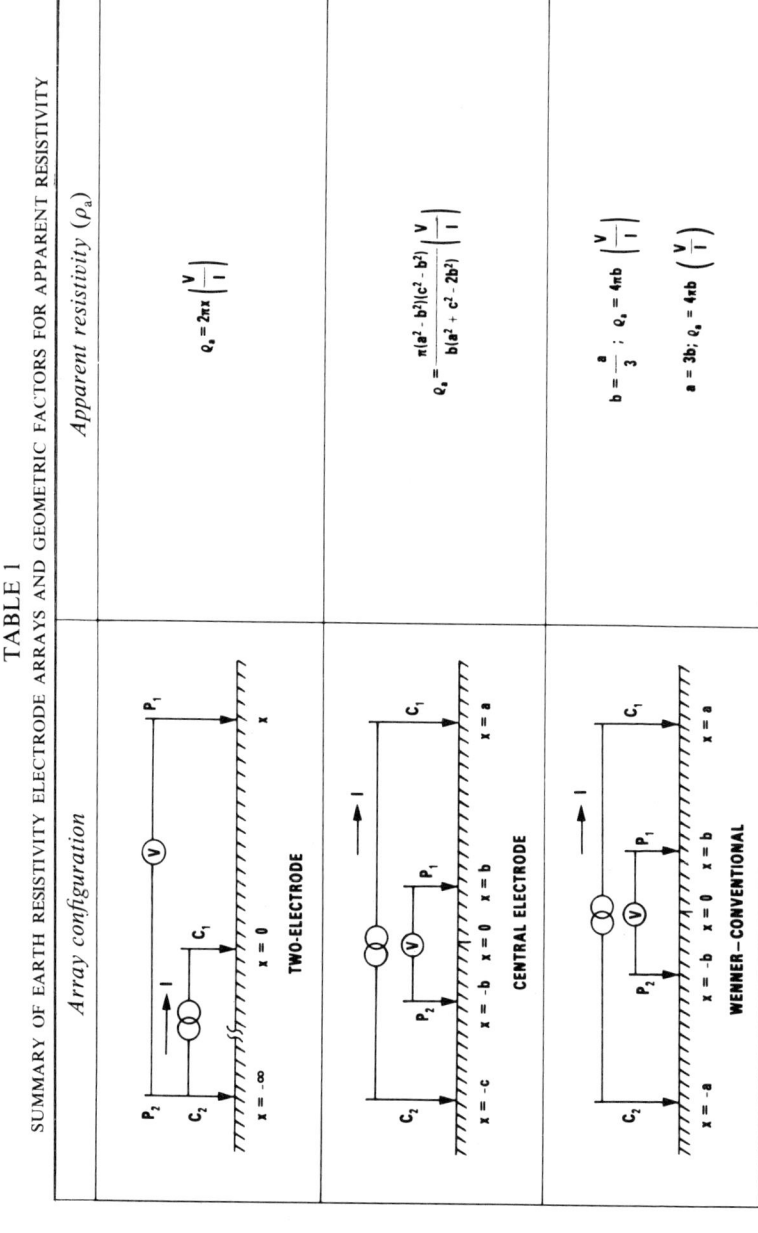

DETECTION AND MAPPING OF TUNNELS AND CAVES

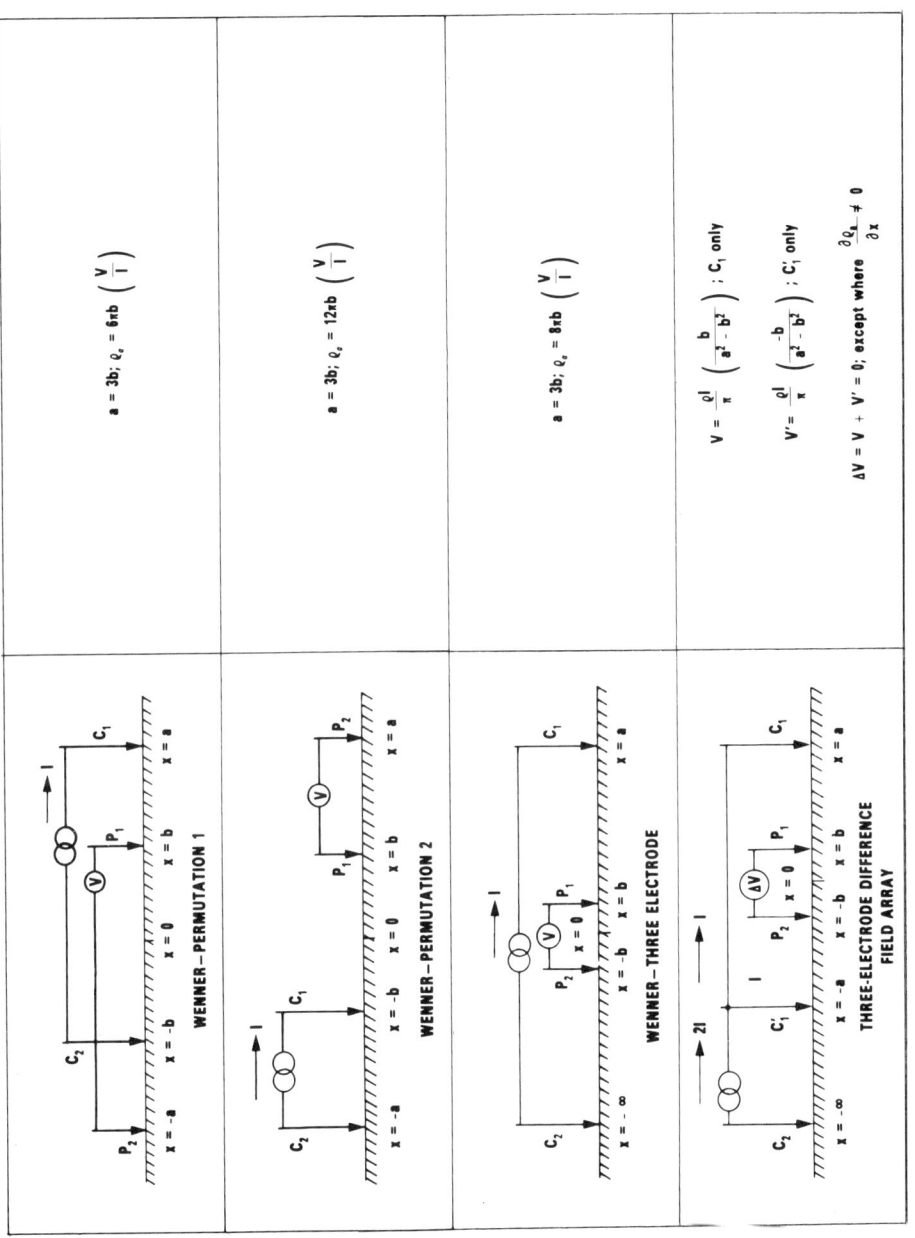

TABLE 1—contd.

Array configuration	Apparent resistivity (ρ_a)
SCHLUMBERGER–CONVENTIONAL	$\rho_a = \dfrac{\pi(a^2 - b^2)}{2b} \left(\dfrac{V}{I}\right)$ $b \ll a: \rho_a \cong \dfrac{\pi a^2}{2b} \left(\dfrac{V}{I}\right)$
SCHLUMBERGER–PERMUTATION 1	$\rho_a = \dfrac{4\pi ab(a - b)}{(b^2 + 4ab - a^2)} \left(\dfrac{V}{I}\right)$
SCHLUMBERGER–PERMUTATION 2	$\rho_a = \dfrac{4\pi ab(a + b)}{(a - b)^2} \left(\dfrac{V}{I}\right)$

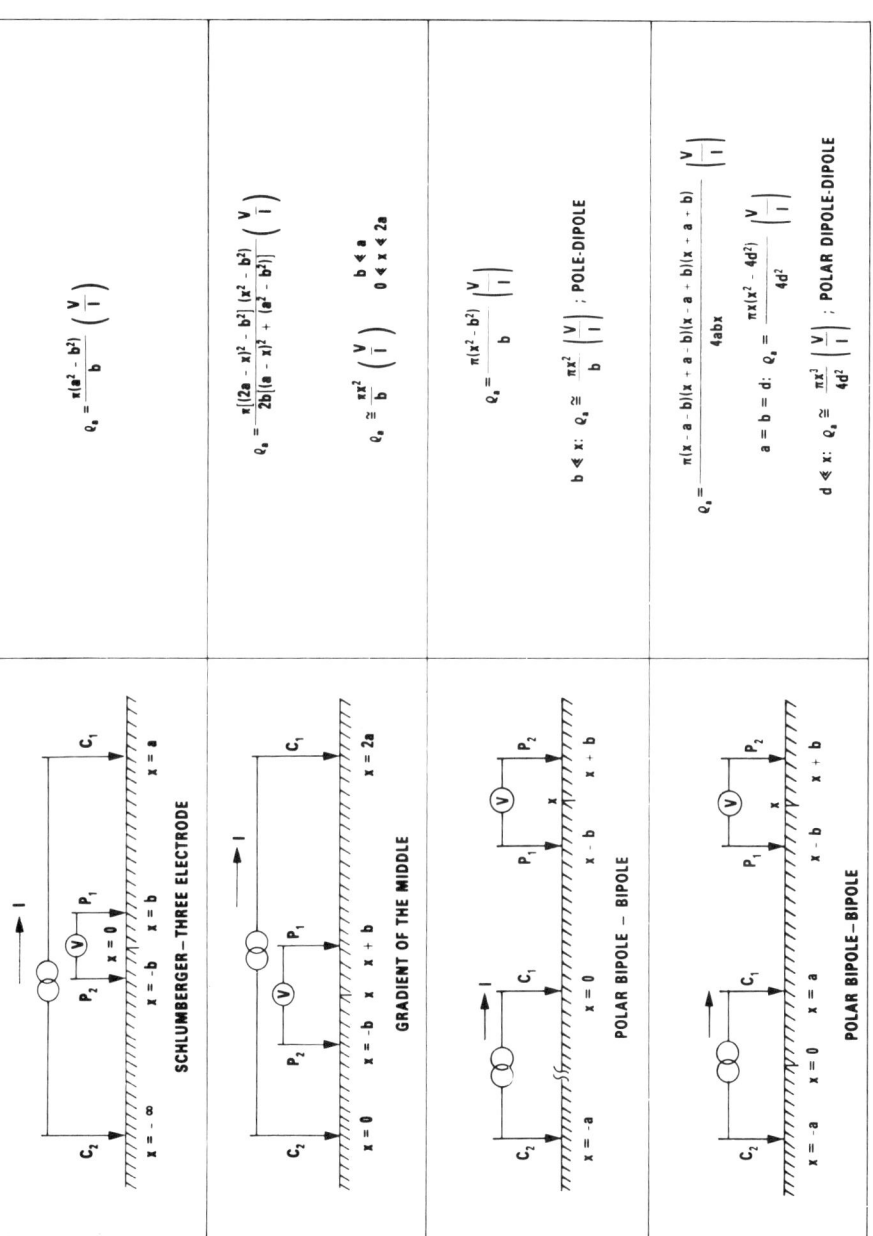

TABLE 1—contd.

Array configuration	Apparent resistivity (ρ_a)
PARALLEL BIPOLE – BIPOLE (diagram with C_1, C_2, P_1, P_2, $x = -a$, $x = 0$, $x = a$, angle θ, R, points b, b)	$\rho_a = \dfrac{2\pi\left(\dfrac{V}{I}\right)}{\left[\dfrac{1}{r_1} - \dfrac{1}{r_2} - \dfrac{1}{r_3} + \dfrac{1}{r_4}\right]}$; $r_1 = \sqrt{(x-a-b)^2 + y^2}$ $r_2 = \sqrt{(x+a-b)^2 + y^2}$ $r_3 = \sqrt{(x-a+b)^2 + y^2}$ $r_4 = \sqrt{(x+a+b)^2 + y^2}$ $R = \sqrt{x^2 + y^2}$ $\theta = \mathrm{TAN}^{-1}\left(\dfrac{y}{x}\right)$
EQUATORIAL BIPOLE – BIPOLE (diagram with C_1, C_2, P_1, P_2, $x = -a$, $x = 0$, $x = a$, $\theta = 90°$, $y = R$, points b, b)	$\rho_a = \dfrac{2\pi\left(\dfrac{V}{I}\right)}{\left[\dfrac{1}{2r_1} - \dfrac{1}{2r_2}\right]}$; $r_1 = r_4 = \sqrt{(a-b)^2 + y^2}$ $r_2 = r_3 = \sqrt{(a+b)^2 + y^2}$ $a = b = d$: $\rho_a = \dfrac{4\pi y \sqrt{y^2 + 4d^2}}{\left[\sqrt{y^2 + 4d^2} - y\right]}\left(\dfrac{V}{I}\right)$ $d \ll y$: $\rho_a \approx \dfrac{2\pi y^3}{d^2}\left(\dfrac{V}{I}\right)$; EQUATORIAL DIPOLE-DIPOLE

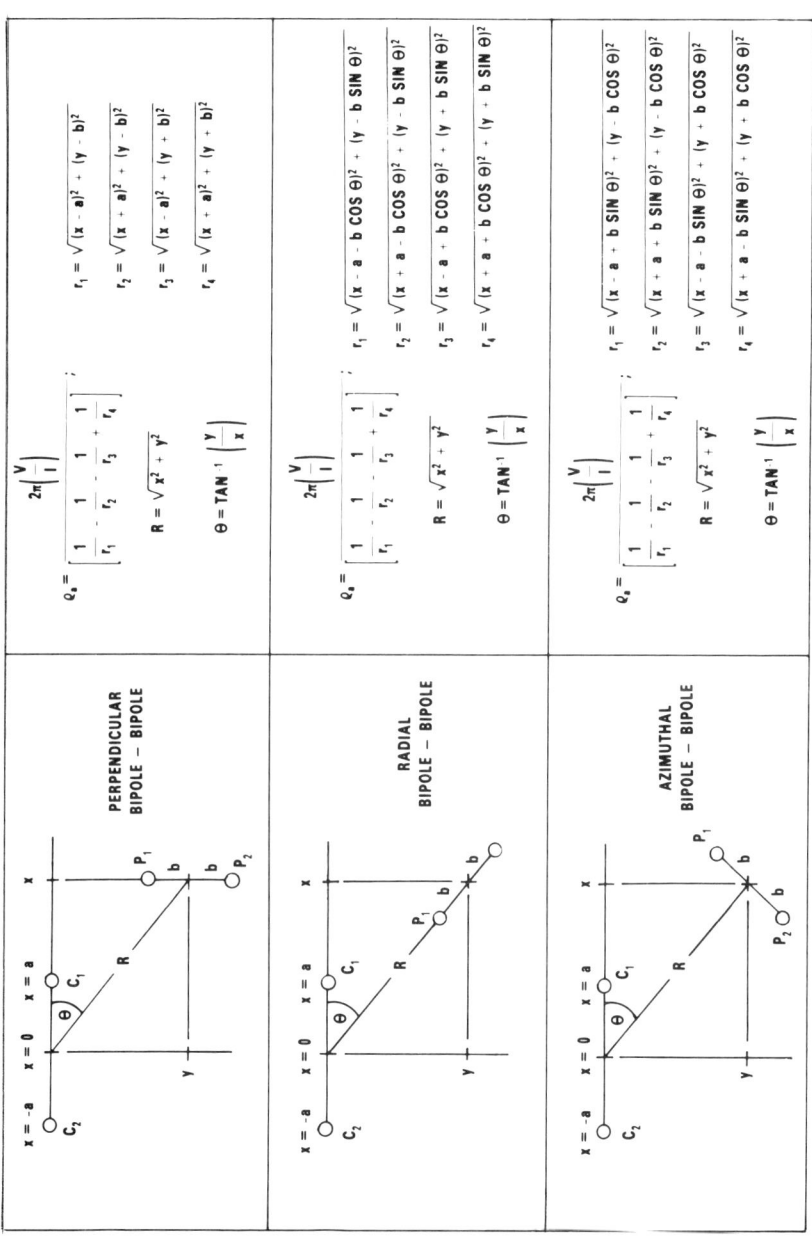

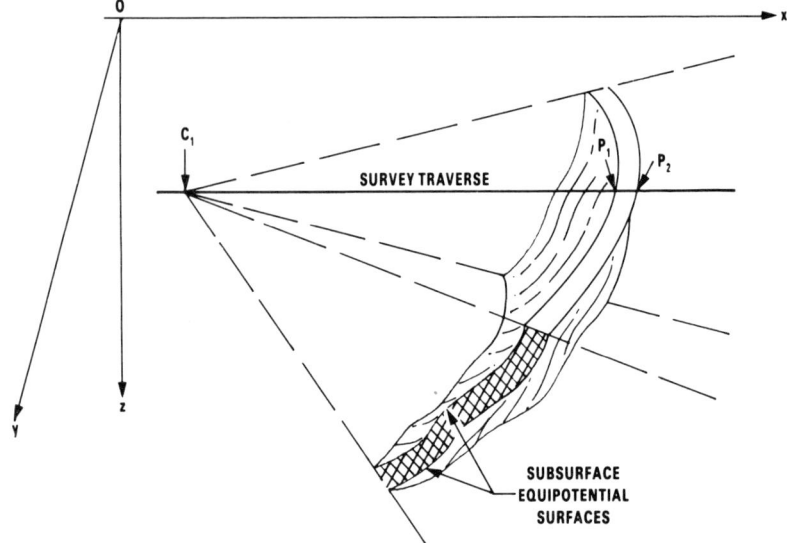

FIG. 2. Simplified resistivity response zone for the pole–dipole electrode array.

arrays because of its relatively large electrode separation and, hence, the greater volume of ground encompassed in its measurements.

For those array configurations such as the pole–dipole in which the potential measurement electrodes are relatively close together, the equipotential lines at each electrode contact on the surface may be considered as the edges of curved equipotential surfaces extended below the surface. Figure 2 illustrates this concept in a simplified way, showing the restricted subsurface region in which an underground anomaly might have influence on the apparent resistivity as determined from the positions of electrodes C_1, P_1 and P_2. The effective measurement volume of this underground region also increases with depth in this case, but only at a fraction of that of the Wenner array or other wide electrode configurations. Thus, in simplified terms of comparison, the Wenner array may be viewed as measuring the resistivity contrast of a given anomaly in the complete volume of an approximate hemispherical region whereas the pole–dipole array measures the anomaly contrast within only part of a spherical shell volume. Obviously, the same anomaly will appear more pronounced in the pole–dipole response than in the Wenner response.

When the earth medium containing the underground anomaly is assumed to be a flat homogeneous halfspace of uniform resistivity, the shell segment shown in Fig. 2 is precisely spherical so that circular arcs centred on electrode C_1 define its curvatures. The graphic method of detecting cavities developed by Bristow[5] is based upon this simplified concept and has proven to be effective in cave and tunnel detection. Both the lateral response and the depth response along the survey traverse is limited to an angular sector of relatively narrow extent. Bristow reported the surface sector angle to be in the range of about 25° and his depth interpretations often extended to depression angles of about 60–70°. More recent work applied to interpretation of weaker cavity anomalies indicate that both the lateral and depth angles of view do not exceed about 45° in relatively homogeneous ground.[12]

As implied by Fig. 2, a cavity or other localised anomaly will be detected if it intersects the equipotential surfaces. However, its location within the volume of the shell is indeterminant without the benefit of other information. This supplementary information is gained by devising spatial survey techniques whereby the anomaly response is observed from a number of independent vantage points determined by the electrode locations and spacings. In the chronology of the various resistivity survey methods applied to cavities, Palmer employed an array of the Schlumberger type in a fixed-station resistivity depth-sounding procedure (somehow known to be located directly over the cavity) in which the electrode geometry factor (b/a) is held constant as the dimension, a, is increased to a limit greater than the expected cavity depth. The interpretation of anomaly depth, h, in this case is determined by the relationship,* $h = a_0 \sqrt{b/a}$, in which a_0 is the distance from the sounding station to the current electrodes for the condition where the maximum anomalous perturbation in apparent resistivity is observed. Typical values of b/a used by Palmer were in the range of 0·25–0·64.

The pole–dipole survey technique used by Bristow employs depth sounding profile measurements in which the potential electrode pair is moved incrementally away from the current source station, first in one

---

* The theoretical basis for this relationship is that of a spherical air-filled cavity and its first-order image influence in a uniformly resistive flat halfspace. An additional relationship of interest is the approximate value of the maximum perturbation in apparent resistivity for various spherical cavity sizes of radius, R

$$\left(\frac{\rho_a}{\rho}\right)_{max} = 1 + \frac{(1-(b/a))}{(1+(b/a))^2}\left(\frac{R}{h}\right)^3$$

indicating the desirability for b/a to be small and the cavity to be relatively large in comparison with its depth.[2]

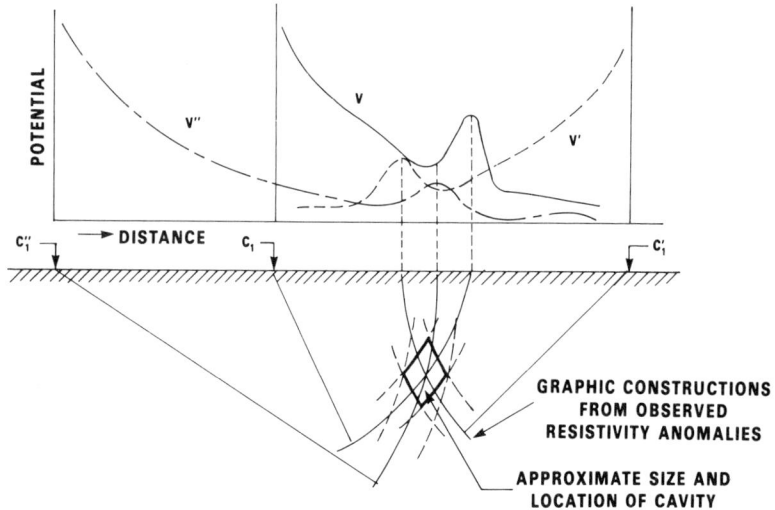

FIG. 3. Graphical analysis technique used to locate the position and depth of cavities in homogeneous ground.

direction and then in the opposite direction along the traverse. Then, by moving the current source location along the traverse at incremental distances so that the measured resistivity profiles will overlap, the intersection of two or more equipotential hemispherical shells having radii corresponding to the current-to-potential electrode separation distances at which resistivity anomalies are observed will locate the subsurface cavity. When this method is applied with sufficient overlap of the resistivity profiles, the subsurface zone of intersection can provide a reasonably good indication of the cavity target cross-sectional size and depth. A typical survey layout used by Bristow employed a fixed potential dipole spacing of 20 ft scanned at 10-ft intervals along the traverse to maximum distances of 200 ft on each side of each current station. Current stations were separated 150 ft apart for general reconnaissance surveys and at 50-ft intervals for detailed surveys. Figure 3 shows the graphical method of locating a cavity using circular arcs drawn from high resistivity anomalies or associated dipole potentials observed in two separate sounding profiles overlapping the target.

Bates[6] adapted the pole–dipole technique devised by Bristow to search for smaller solution caves and grike zones (soil-filled erosion channels in the top of limestone bedrock) in karst terrain. For that purpose, the potential electrode spacing was reduced from 20 ft to 10 ft to provide finer resolution and the current stations were spaced at 50-ft intervals to provide greater overlap in the measured resistivity profiles. Thus, by requiring a minimum of three or four intersections of the anomaly perturbation arcs at a common underground position, a much higher confidence is gained in detecting and locating relatively small cavity targets. Using this method and a graphical analysis designed to localise targets having either high or low perturbations in apparent resistivity, Bates reported that low resistivity grikes could be readily located and, in one case, an 8-ft diameter air-filled solution cave was detected at a depth of 120 ft below surface.

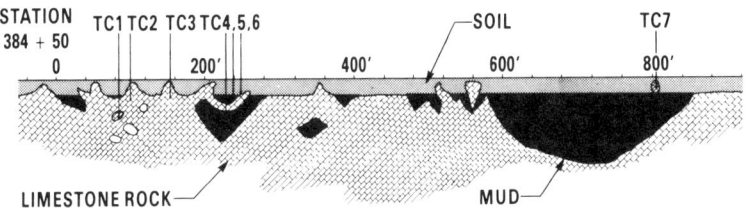

FIG. 4. Vertical profile under Survey Traverse A, Interstate Highway 59 test site, Birmingham, Alabama.[7]

The successful results achieved by Bates indicated that the pole–dipole method was potentially applicable to the problem of locating shallow cavities in soil associated with sinkhole formation and underground mud flows in karst terrain. This application was explored by the Southwest Research Institute[7] with good results in detecting mud flow zones under highways and in detecting small air-filled cavities and piping in soils within a 30-ft thick surface soil layer overlying highly porous limestone bedrock. Test borings at the interpreted anomaly locations confirmed the presence of predicted fluidised soil zones at depths of 25–30 ft and several air-filled cavities having diameters as small as 3–4 ft. The field techniques employed in these surveys were essentially equivalent to those used by Bates with the exception that the current station intervals were reduced from 50 ft to 20 ft to provide more redundancy in the overlapping resistivity profiles and better coverage of the shallower depths of interest.

Figure 4 shows the final results of one of several 1000-ft pole–dipole surveys along a four-lane highway where soil sink instabilities were suspected. Depicted in this interpretation are four small cavities in limestone, numerous limestone pinnacles or ridges in the soil overburden, soil and mud-filled grikes in the limestone bedrock, two mud-filled cavities and a large traverse mud flow. The soil layer is approximately 32-ft thick. The limestone ridges were found as predicted and verified by test cores TC2 and TC3 to be within 5–6 ft of the surface. The high-resistivity cavities were estimated to be about 10–15 ft in diameter at a depth of 45–65 ft. Test core TC1 verified one such cavity containing highly fractured limestone at a depth of 40 ft. The interpreted mud-filled cavity shown in Fig. 4 underlying a thin layer of limestone was verified to have a maximum depth of 55 ft by test core TC4. The large mud flow was detected on several other parallel traverses including that shown in Fig. 4. Test corings on an adjacent traverse confirmed the mud flow channel to be about 100 ft deep and over 200 ft wide.

Other electrode arrays have also been considered for the detection of cavities. Habberjam[13] experimentally evaluated the use of a modified Wenner array in which three different pairings of the electrodes were employed for current injection and potential measurements within the same geometrical layout to detect spherical cavities. Model experiments in a brine tank using each electrode arrangement resulted in numerous resistivity anomaly curves for parametric values of electrode spacing, sphere size and sphere depth. These results were then normalised to produce a family of interpretation curves which, for a homogeneous host medium, will permit the size and depth of a suspected spherical cavity to be determined. For deepest detection, these interpretation curves indicate that the equispaced electrode separation distance should be in the range of one to two times the radius of the sphere. Further, if the geological noise in the field survey data is 5–10%, then the limit of detection of the sphere will be in the depth-to-diameter range of about 0·5–0·75. Van Nostrand[14] derived a similar limit of detection for a perfectly conducting sphere based upon the use of a conventional Wenner electrode array.

Comparative analysis of several surface electrode arrays have been investigated by Militzer et al.[15] including the permutated Wenner arrangement described by Habberjam, using a two-dimensional theoretical analysis (i.e. an infinite tunnel cavity and infinite line electrodes) as a basis for relative performance comparison. In addition to the Schlumberger, Wenner pole–dipole and dipole–dipole arrays, Militzer et al.[15] evaluated three additional arrays including a gradient of the middle array (similar to Schlumberger array but with moveable potential electrode pair and current electrodes held fixed), a three-electrode difference field array (similar to pole–dipole in which current is injected separately at symmetrical locations on each end of the potential pair and the difference in potential readings used to characterise the anomaly), and a focusing array[16] in which two equal source currents are simultaneously injected at equal distances outside the ends of the potential pair in the same manner as the guard currents employed in a focusing electric well-logging probe.

Militzer et al.[15] have computed a relatively large range of master curves for the arrays mentioned above and have developed an 'Anomaly Effect' (AE) parameter for comparative evaluation of cylindrical target detection in a homogeneous halfspace. The Anomaly Effect is defined in terms of the extreme values of normalised apparent resistivity

$$\frac{\rho_a}{\rho_0} = 1 + \frac{V_d}{V_0} \tag{3}$$

where V_d = potential difference resulting from target anomaly;
V_0 = potential difference observed in the absence of the target anomaly;
ρ_0 = resistivity of the host medium.

In particular, the Anomaly Effect is

$$AE = \left(\frac{\rho_a}{\rho_0}\right)_{max} - \left(\frac{\rho_a}{\rho_0}\right)_{min} \tag{4}$$

corresponding essentially to the peak-to-peak variation in the target anomaly response without the masking effects of the normal response, V_0, of the host medium.

Table 2 is a very generalised summary of the theoretically derived Anomaly Effect for six of the electrode arrays considered. Target responses having $AE \geq 0.1$ for various depths below surface are indicated by a + sign implying successful detection in a 10% geological noise background. The Wenner and Schlumberger arrays appear to have essentially equivalent performance, with a target detection limit of about $h/R = 3$ (i.e. a depth-to-diameter ratio of 1·5). The three-electrode array (Wenner or Schlumberger with current electrode $\to \infty$) and dipole–dipole and mid-gradient of the middle arrays have increasingly better performance to a target depth-to-diameter limit of 2. Finally, the three-electrode difference field array provides a successful detection limit of $h/R = 5$ corresponding to a target depth-to-diameter ratio of 2·5.

The performance of a focused array is depicted in Fig. 5 in which the 'Cavity Effect' (CE) is defined simply as

$$CE = \left(\frac{\rho_a}{\rho_0}\right)_{max} - 1 = \left(\frac{V_d}{V_0}\right)_{max} \tag{5}$$

corresponding to the peak perturbation in apparent resistivity with and without the cylindrical tunnel target present. In this case, the focusing electrode array appears to provide a successful detection limit at target

TABLE 2
RESISTIVITY ANOMALY DETECTION LIMITS FOR VARIOUS ELECTRODE ARRAYS[15]

ARRAY CONFIGURATION	COMPARATIVE PERFORMANCE			
	$\frac{h}{R}=2$	$\frac{h}{R}=3$	$\frac{h}{R}=4$	$\frac{h}{R}=5$
WENNER – CONVENTIONAL	+	+	–	–
– PERMUTATION 1	+	+	–	–
– PERMUTATION 2	+	+	+	–
– THREE ELECTRODE	+	+	+ (?)	–
SCHLUMBERGER – CONVENTIONAL	+	+	+ (?)	–
– PERMUTATION 1	+	+	–	–
– PERMUTATION 2	+	+	–	–
– THREE ELECTRODE	+	+	+	–
DIPOLE – DIPOLE	+	+	+	+ (?)
GRADIENT OF THE MIDDLE	+	+	+	+ (?)
THREE-ELECTRODE DIFFERENCE FIELD ARRAY WENNER	+	+	+	+
SCHLUMBERGER	+	+	+	+

h = Depth } of buried cylinder target
R = Radius
+ = Anomaly Effect ≥ 0.1
– = Anomaly Effect < 0.1

depths of about twice that obtained with the best performing array listed in Table 2. From practical field tests, however, Militzer et al.[15] show that the focusing array, while it is more effective than the three-electrode difference array, provides an anomaly enhancement of only about 25–30% rather than the 100% improvement implied in Fig. 5.

The work of Militzer et al.[15] is significant in the progress of developmental work in geoelectrical methods of cavity detection for two reasons. First, in much of their work they advocate the direct use of the measured surface potentials rather than the apparent resistivity derived from the potentials and geometric factors. This recommendation may be extended to imply that the target anomaly may be stronger and more distinctively recognisable in terms of its disturbance in otherwise predictable potentials for the host medium alone rather than in its influences on apparent resistivity. The variation from standard earth resistivity interpretation practice as represented by their Anomaly and Cavity Effect parameters for localised targets such as cavities and tunnels is also worthy of note.

The second point of importance in the work of Militzer et al.[15] is

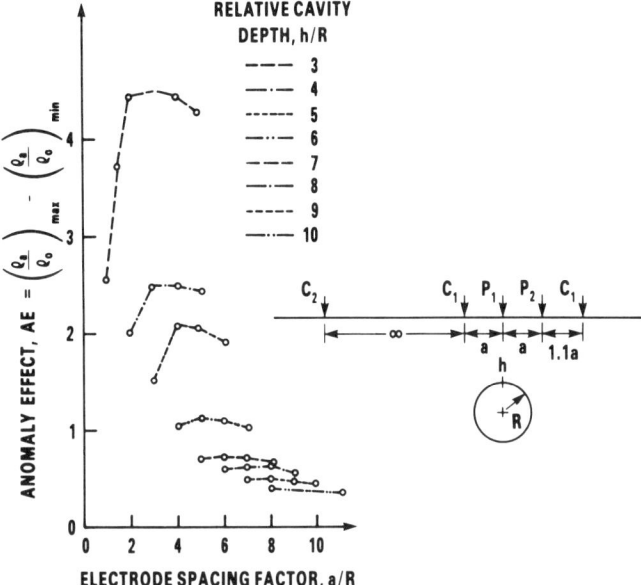

FIG. 5. Anomaly effect of two-dimensional target using focusing array.[15]

complementary to the first in that the best performing electrode arrays are those in which the normal response of the host medium is reduced. A clear example of this is the three-electrode difference technique in which only the secondary fields from the target contribute to the anomalous response. Focusing arrays of appropriate geometry also reduce the normal response of the host medium, thereby providing a basis for improved target detection signal-to-noise ratio.

3.1. Advanced Resistivity Techniques

The preceding earth resistivity survey methods of cavity detection utilise manual field procedures for data acquisition and graphical methods of data display and interpretation. The necessity for close electrode spacings in achieving good resolution of small targets and the need for overlapping depth soundings extend these manual methods to their practical limits in terms of field time and manpower. Recently, however, this problem has been substantially overcome through the development of an automatic earth resistivity data acquisition system.[10,11]

The need for rapid high-resolution earth resistivity surveys was first dictated by military requirements for detecting subversive intrusion tunnels. An advanced Earth Resistivity Data Acquisition System (ERDAS) was developed to automate the manual pole–dipole technique employed earlier by the Southwest Research Institute for detecting and mapping soil sinkholes, limestone solution caves and tunnels. Efficient automation of the pole–dipole survey technique required a reversal of the manual survey procedure whereby each current electrode station is usually held at a fixed position to inject a constant current while the potential electrode pair is moved to obtain the associated voltage readings. In contrast, the automatic technique sequentially injects the constant current into an array of current electrode positions while the potential pair is held at a fixed position to measure the associated voltages. Thus, by laying out all of the current electrodes along the survey traverse and providing a means for remote current switching into each electrode, in sequence, the voltages at a given location along the traverse corresponding to each current source station are automatically measured and recorded.

After all current stations have been excited in sequence, the potential pair is moved one dipole increment along the traverse and the current injection sequence is repeated to yield another sequence of voltages. By repeating this procedure until the potential pair has been moved along the entire traverse line, the same field data will have been obtained as that acquired using the normal field procedure. Data recording is accomplished in digital form on magnetic tape and includes information on the current and potential electrode positions, the injected current and the potential reading for each sequential step of current injection along the traverse. Current and voltage values are measured and recorded within an accuracy of $\pm 0.1\%$. The ERDAS equipment employs 64 current injection electrodes.

Each sequential measurement step requires a time duration of 2 s and, therefore, the voltages at one potential measurement location as produced by 64 current electrode stations will require only 128 s. With the current electrodes separated by 10 m, corresponding to a 630-m traverse length, and with the potential electrodes separated by 2 m and moved in 2-m increments along the traverse, all possible combinations of current and potential pair positions (20 480) can be recorded in about 17 working hours corresponding to nearly 1500 recorded readings/h. However, since depth soundings to about 100 m at a given current station do not require the current to be injected at all 64 source electrode positions on every sequence, a practical automatic survey time for a 630-m traverse length may be reduced to about 9 h, not including electrode layout and retrieval. The automatic survey technique can accomplish in approximately 9 h a survey which would require about 9 days using manual survey procedures.

Figure 6 illustrates the automatic resistivity survey method. The ERDAS

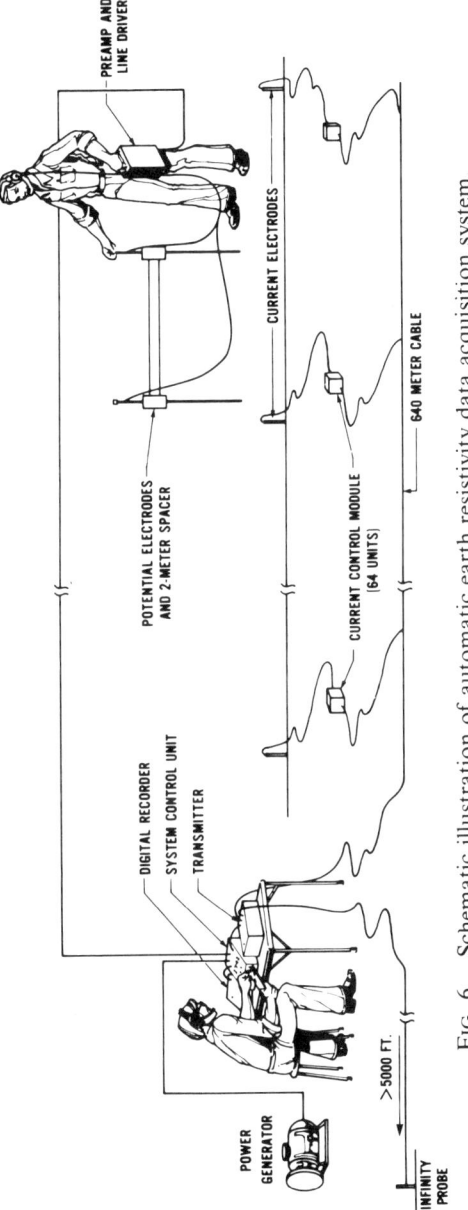

FIG. 6. Schematic illustration of automatic earth resistivity data acquisition system.

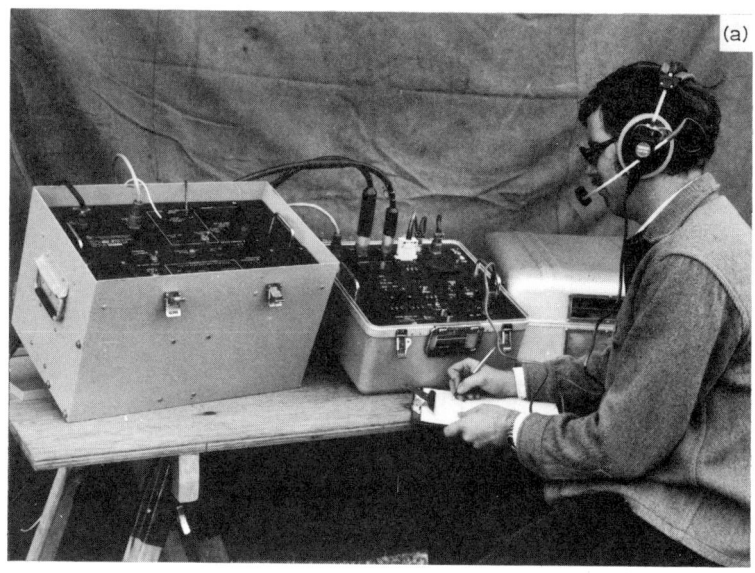

Fig. 7. Earth Resistivity data acquisition system: (a) primary operating station; (b) potential data telemetry cable and carrying hasp.

FIG. 7—contd. (c) Current control module and ground electrode; (d) belt-mounted potential measurement preamplifier.

FIG. 8. Potential measurement electrode array (2-m spacing) and remote operator.

equipment can provide a selectable constant current in the range of 0·1–2 A with a maximum supply voltage of up to 800 V. The current is low-frequency AC having a selectable frequency in the range of 10–25 Hz. Low-noise potential measurements are accomplished by means of a high-impedance balanced-input preamplifier having a selective frequency response and automatic gain ranging. The preamplifier and its gain-ranging capability permit voltage measurements over a 120 dB dynamic range with an accuracy and resolution of ±0·1% or better. Measured values of potential are compensated for small changes in the constant current delivered to each electrode so that each voltage reading is precisely that which would be obtained if a perfect current source were driving each electrode. The voltage readings, preamplifier gain and electrode positions for each reading are recorded on cassette magnetic tape. The accuracy and dynamic range of the potential data are preserved by use of 12-bit digital resolution together with 0–40 dB preamplifier gain range control.

In operation, the primary system operator will first dial in the potential pair location along the 630-m long traverse, dial in the starting and stopping current station electrodes, and then start the automatic data acquisition process. When the programmed current sequence is completed he will instruct the operator at the potential measurement location via a voice-communication line contained in the voltage data transfer cable to move the electrode pair one increment. Then, readjusting the potential-electrode position thumbwheel dials and current starting and stopping thumbwheel dials, the primary operator will initiate the next current injection sequence. In this manner, the entire traverse length may be surveyed. Additionally, if a survey longer than 640 m is desired, the ERDAS system is designed to permit the 80-m long segments of current cable and voltage data transfer cable located at the rear of the traverse line to be moved to the forward end of the traverse so as to permit long continuous surveys of indefinite length.

Figures 7 and 8 show the major elements of the ERDAS equipment in field use. Figure 7(a) shows the primary operating station with the constant current generator, the system control unit and the magnetic tape data recorder arranged from left to right. Figure 7(b) shows a typical 80-m length of cable and its method of man-pack transport. Figure 7(c) shows a typical current electrode station with its digitally addressed current control module connected to a moulded takeout on the current cable. Figure 7(d) shows a portable potential measurement preamplifier. The photograph in Fig. 8 shows the potential electrode assembly with its 2-m fibreglass electrode separator installed in rough ground and the preamplifier and calibrator unit mounted on the operator's belt.

3.2. Characteristic Resistivity Responses from Cavities

The surface potential distributions or apparent resistivity signatures associated with localised subsurface resistivity contrasts in an otherwise homogeneous medium have been studied by several investigators in the past. These analyses may be subdivided into categories based upon subsurface target contrasts and geometrical shape. For example, spherical resistivity anomalies have been analysed to include both perfectly-conducting bodies[14,19] and arbitrary resistivity contrasts.[17,18,20] Spheroidal and ellipsoidal[21,22] shapes have been examined from the viewpoint of using surface electrodes or buried electrodes (in boreholes) or a combination of surface and buried electrodes. Cylindrical[15] shapes and various prismoidal[12] cross-sections of finite length have also been studied.

In many of these analyses where the results are not specialised to unusual electrode arrays, the characteristic responses are comparable in a general way. For such purposes of comparison, Fig. 9 presents reproductions of several selected signature characteristics representative of localised subsurface targets, including finite and infinite resistive contrasts with the host medium. As may be noted from these various responses, the general features are similar; namely, the apparent resistivity varies both above and below that of its homogeneous host medium with the maximum variation

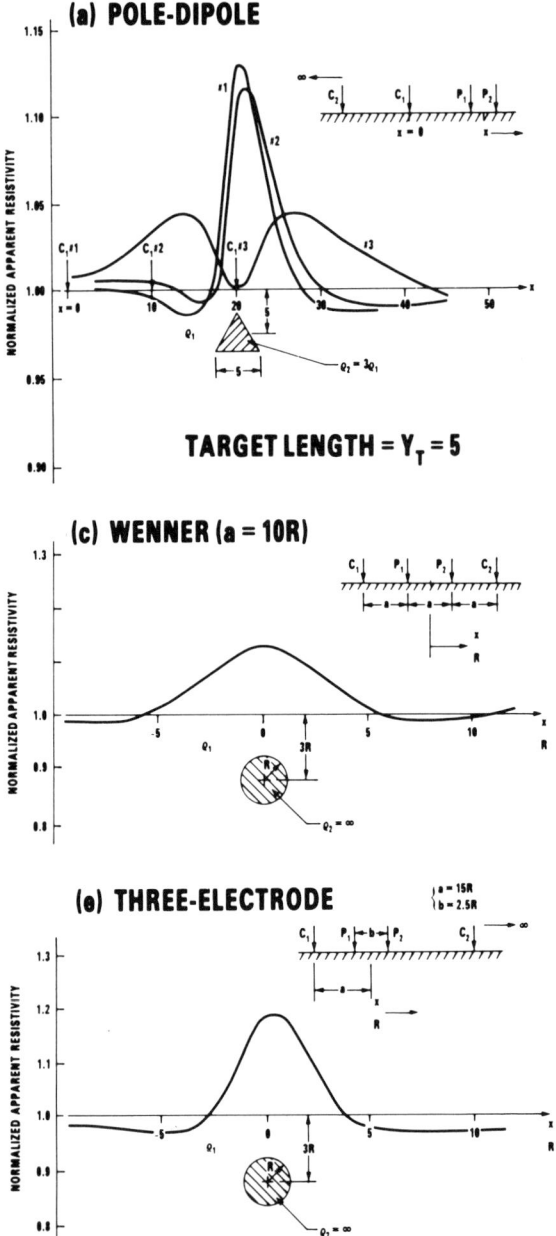

FIG. 9. Normalised resistivity anomalies from various cavity targets and electrode arrays. Figures (a) and (b) are 3-D responses to finite-length targets.[12] Figures (c)–(f) are 2-D responses for infinite tunnel targets.[15]

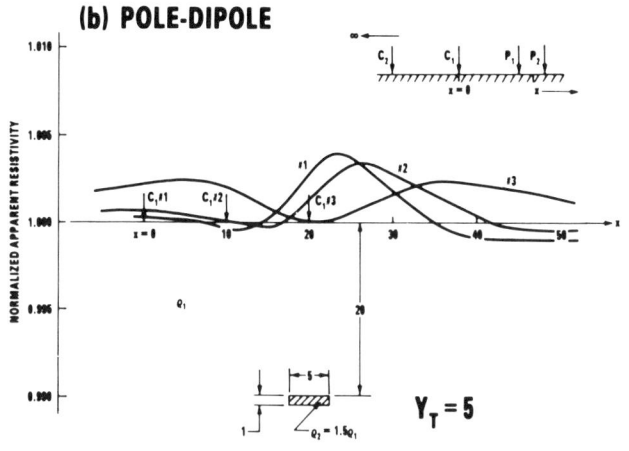

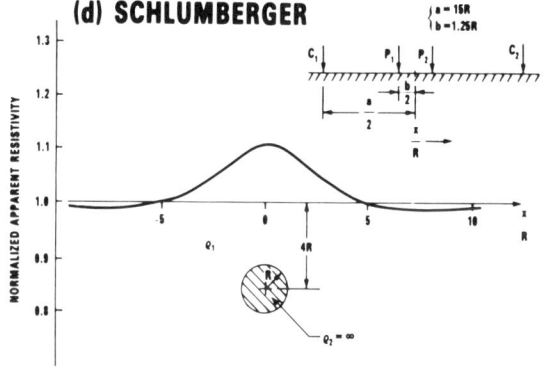

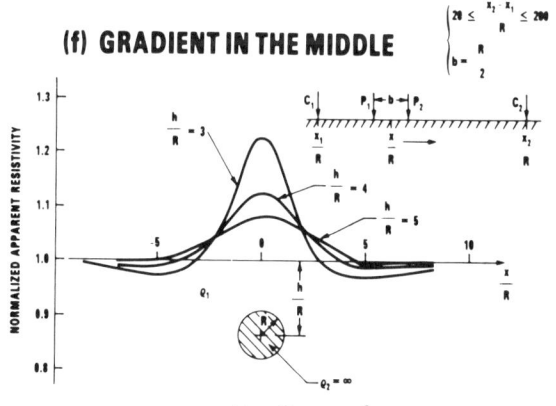

FIG. 9—contd.

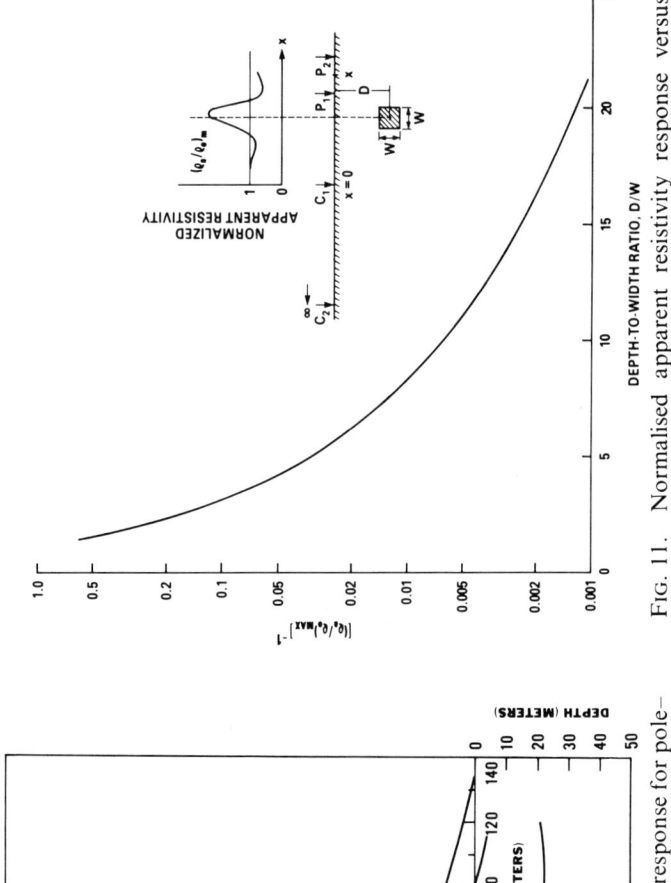

FIG. 10. Normalised apparent resistivity response for pole–dipole measurement over long tunnels (2×2 m square cross-section).[12]

FIG. 11. Normalised apparent resistivity response versus depth-to-width ratio for pole–dipole measurements over square cross-section tunnels.

being a function of the contrast between the target and host medium. The peak value of the anomaly is generally over the target. Other features of the responses relate to the target depth. In particular, the maximum amplitude of the response is very sensitive to the depth-to-diameter ratio of the target as is the spatial width of the anomalous amplitude peak.

Figure 10 illustrates the pole–dipole response characteristics of a 2 × 2-m square tunnel located at different depths below the surface. These results yield an approximate empirical response sensitivity for a point source of current and a square cross-section tunnel expressed by

$$\left(\frac{\rho_a}{\rho_0}\right)_{max} = 1 + 1\cdot6\left(\frac{D}{W}\right)^{-2\cdot4} \qquad (6)$$

where D is the depth to tunnel axis and W is the width of the tunnel. This relationship is presented graphically in Fig. 11 emphasising primarily the relatively deep target case. As indicated, the depth limit at which the perturbation in apparent resistivity is reduced to $0\cdot1\%$ is 43 m for a 2 × 2-m square cross-section tunnel.

The spatial response signature of the tunnel target is dependent upon the position of the current source electrode relative to the target. This effect is shown in Fig. 12 showing that the maximum response is obtained when the current electrode is located at an offset distance of about four tunnel depths away from the tunnel axis, corresponding to an observation depression angle of 15°. The apparent resistivity anomaly observed when the current source is located directly over the tunnel has two symmetrical response peaks with respect to the tunnel axis whose amplitudes are about half the largest detectable maximum.

In general, apparent resistivity anomalies associated with various different subsurface cavities are similar to those illustrated for elongated tunnel shapes. The depth-to-size ratio and the resistivity contrast, either resistive or conductive, are the primary detection sensitivity parameters. Variations in target cross-sectional geometry which do not impose major changes in the depth-to-size ratio will only introduce small differences in the signature. Other factors which also affect the target response are the orientation of elongated cavities with respect to the survey traverse and, of course, practical imhomogeneities (geological noise) in the resistivity structure of the host medium.

As a practical expedient in representing the general apparent resistivity anomaly of a cavity, particularly in connection with tunnel detection and interpretation, a synthetic target model response has been found useful. For this purpose, the spatial pattern produced by a dipole secondary field with

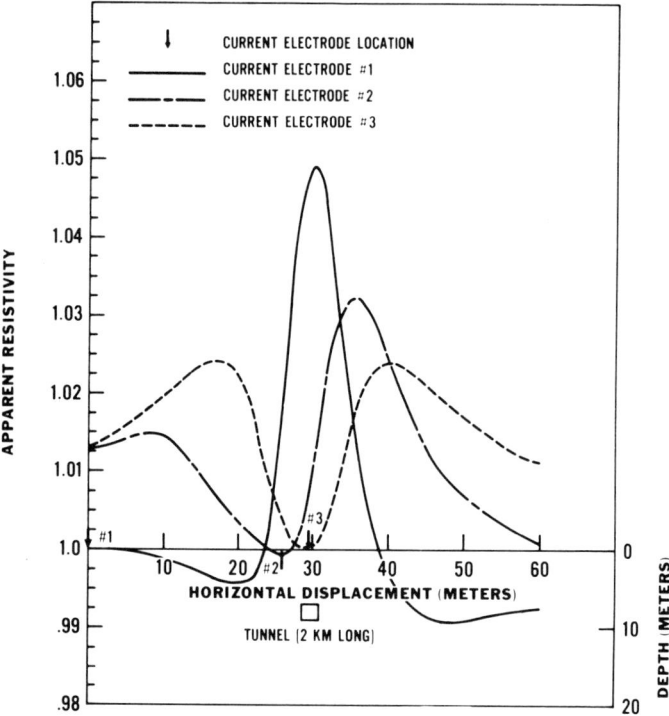

FIG. 12. Normalised apparent resistivity response for pole–dipole measurements at three current electrode locations over long tunnel (2 × 2-m tunnel, 8-m deep).[12]

its polar axis oriented along a line connecting the centre of the cavity and the current station electrode is a realistic cavity response model. Figure 13(a) shows the geometry of this dipole model and its relationship to the current and potential electrodes on the surface.

The simplified representation of apparent resistivity provided by this model may be expressed as

$$\rho_a = \left(\frac{x_1 x_2}{x_2 - x_1}\right)\left(\frac{V_1 - V_2}{I}\right) \qquad (7)$$

where

$$V_1 = \frac{\cos\theta_1}{D_T^2 D_1^2} \qquad V_2 = \frac{\cos\theta_2}{D_T^2 D_2^2}$$

$$D_T^2 = x_T^2 + z_T^2 \qquad D_1^2 = (x_T - x_1)^2 + z_T^2$$

$$D_2^2 = (x_T - x_2)^2 + z_T^2 \qquad \text{(see Fig. 13(a))}$$

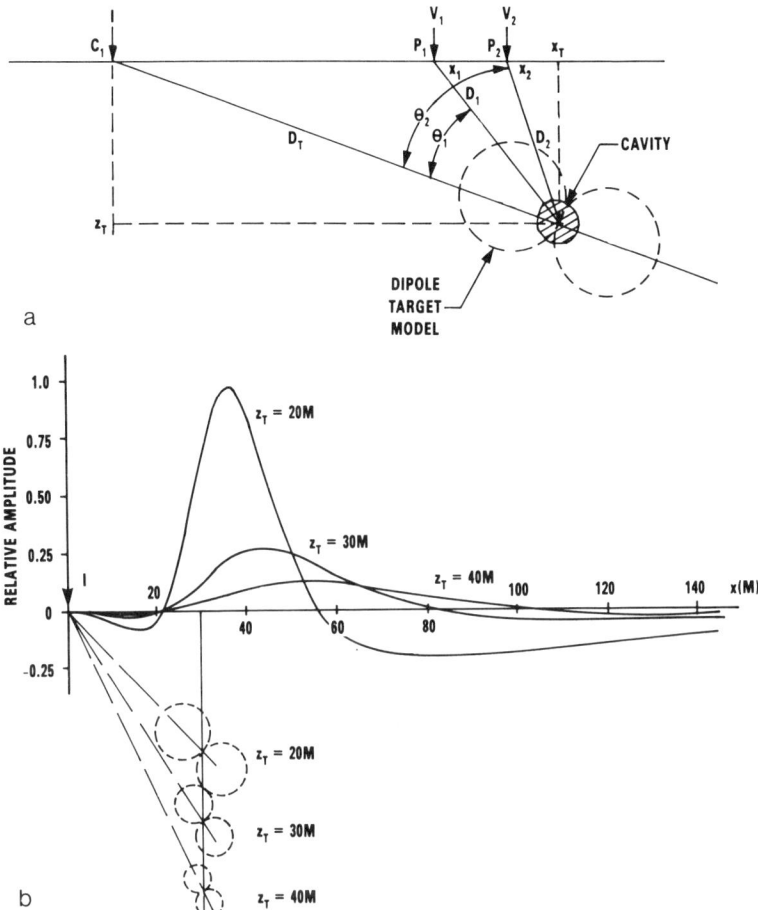

FIG. 13. Synthetic cavity target model and calculated apparent resistivity response. (a) Simplified dipolar cavity target model; (b) relative apparent resistivity response produced by dipolar target model.

The synthesised apparent resistivity anomaly for a cavity target located at 20 m below surface based upon eqn (7) is shown in Fig. 13(b) for the current source electrode 80 m from the target location. This result compares favourably with the anomaly signatures shown in Figs 9–12.

3.3. Data-Processing Techniques

Practical methods used to analyse anomalies in apparent resistivity caused by localised cavity targets are based upon direct interpretation techniques.

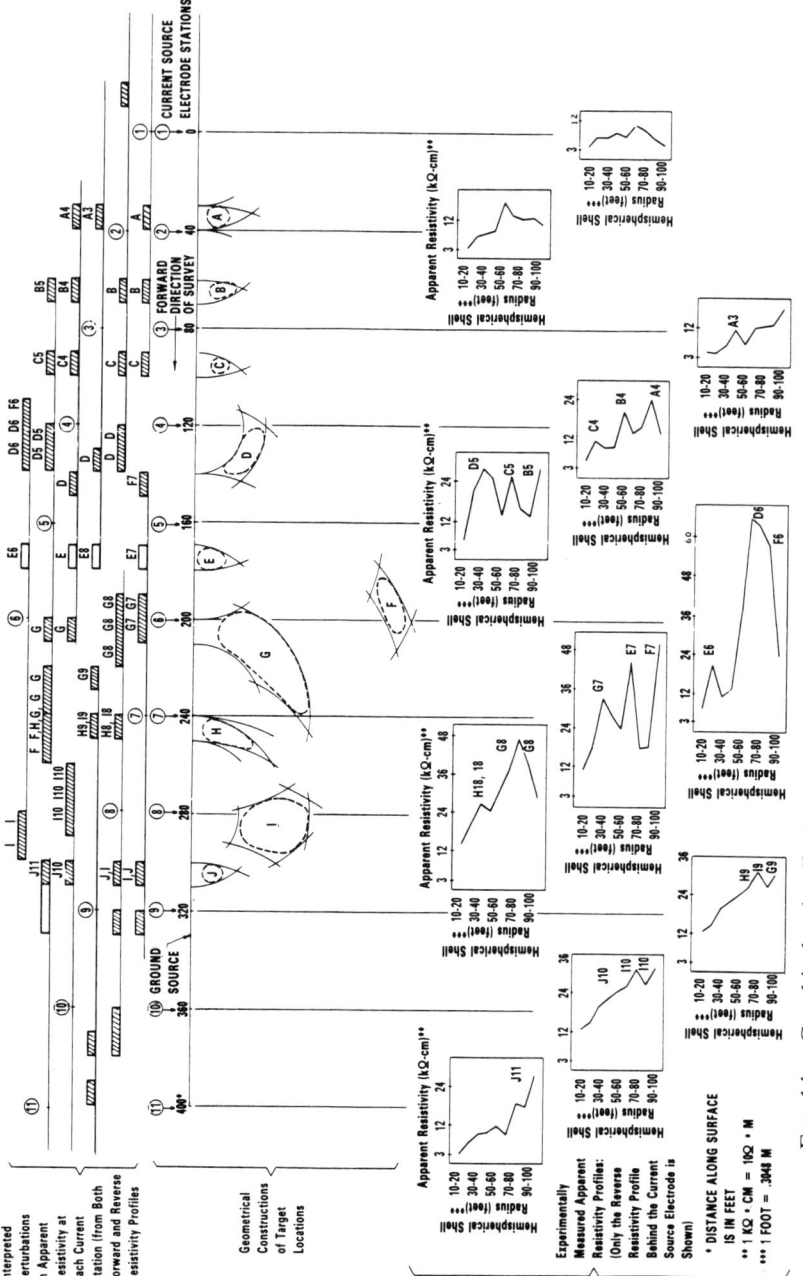

FIG. 14. Graphical pole–dipole resistivity survey over main room of Medford Cave, Florida.[7]

Palmer[2] approached this problem by recognising the characteristic signature of the cavity in a Schlumberger depth-sounding profile centred on the target. With relatively strong target responses, he was able to visually interpret the cavity anomaly in the presence of other larger scale variations in resistivity and, hence, did not require any particularly detailed collateral information on the subsurface resistivity structure of the host ground. However, small scale variations are often introduced into resistivity depth soundings by surface and near-surface conditions thereby masking the response of even fairly shallow targets. Thus, in the presence of geological noise corresponding to variations in apparent resistivity of about 3%, the maximum depth at which a spherical void target can be clearly recognised in a single depth-sounding profile is about 1·5 × the void diameter, i.e. when the overburden thickness is equal to or less than the sphere diameter. Correspondingly, for a square cross-section tunnel target of infinite length the depth at which the perturbation in resistivity is 3% occurs at about 5 × the width of the tunnel.

Target signature recognition under near unity signal-to-noise ratio conditions is generally a marginal process and, therefore, the predictions of detection depth given above are optimistic. Under such threshold detection conditions, there is a good probability that a noise perturbation might be mistaken for a target. To overcome this problem to a useful extent, the interpretation technique devised by Bristow[5] and advanced by Bates[6] and by Fountain[8] using a high resolution pole–dipole array takes advantage of the fact that the geological noise is greatest near the surface and is spatially distributed whereas the cavity target is localised. With this technique, overlapping resistivity profiles can be used to separate noise anomalies near the surface from a cavity target at depth.

Figure 14 shows an example of data interpretation for an overlapping pole–dipole survey over a different kind of target—a shallow limestone solution cavern. Figure 15 shows a map of the cave and three particular survey traverse lines of interest. In these shallow target data, the signal-to-noise ratio of the observable perturbations is very high and the measurements are readily analysed by the graphical method. Circular arcs drawn about each current station at radii corresponding to higher resistance perturbations than the average apparent resistivity of the host medium delineate several high resistance anomalies. The largest anomaly (G, Fig. 14) identified at depth corresponds accurately with the main room of the cave whereas the second anomaly (I) is a lateral response from the nearby section of the generally L-shaped cave. The deepest anomaly (F) is believed to be a small cavity underlying the main room of the cave at about

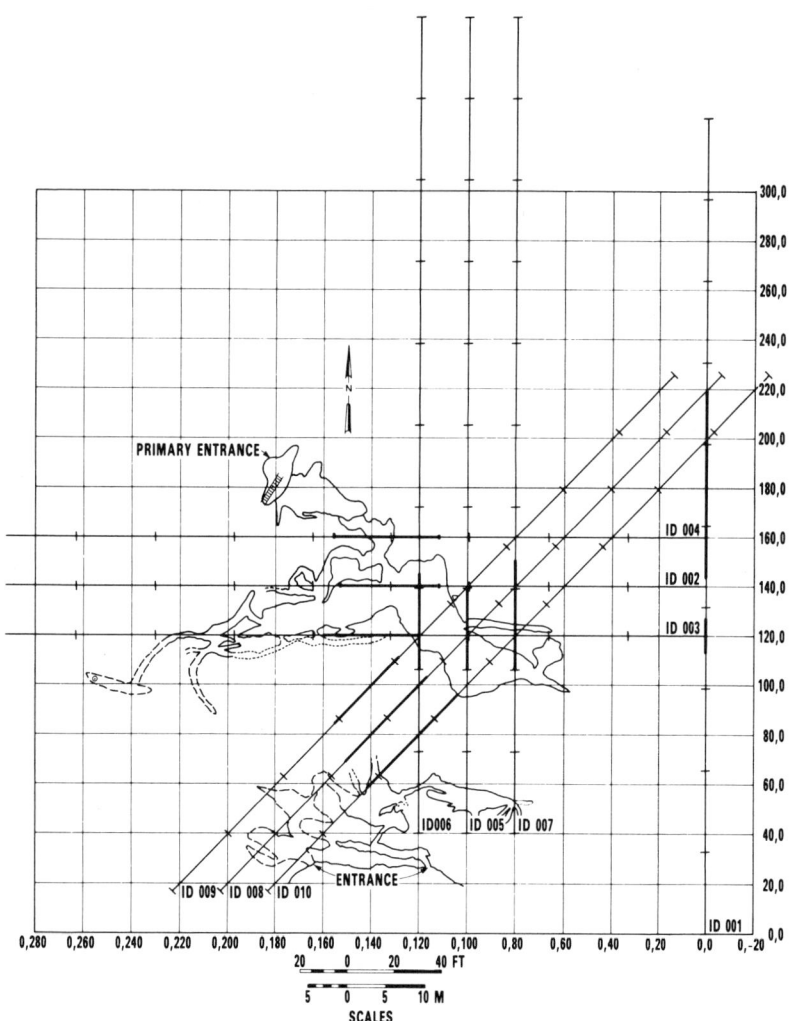

FIG. 15. Map of Medford Cave showing resistivity survey lines.[17]

twice the depth of the main section of the cave. Inaccessible unmapped passages leading downward from the main room were indicative of this possibility; however, test drilling was not practical to verify this interpretation. The relatively small size indicated by this inferred anomaly makes it a target detectable at a depth-to-diameter ratio of about five or six.

The smaller near-surface anomalies (such as indicated by A–E, H, J)

DETECTION AND MAPPING OF TUNNELS AND CAVES 195

were found to be subsurface limestone rock pinnacles in several cases where test borings were drilled. These anomalies tend to exhibit relatively large and well-defined high resistivity perturbations but, they are readily segregated as surface effects. Obviously, by selecting those perturbations which deviate below the average trend of apparent resistivity of the host ground, the same technique of anomaly interpretation can be applied, where appropriate, to targets having a conductive contrast such as mud or water-filled cavities.

Figure 16 depicts a three-dimensional interpretation of the large room of the cave as derived from three parallel traverse lines. This result is in generally good agreement with the actual cave structure and depth below surface; however, because of the increasing resistivity gradient with depth in the host ground, some distortion of the target anomaly with depth must be expected. In general, an increasing resistivity profile with depth will tend to concentrate the injected current nearer the surface and, as a consequence, the depth of an anomaly interpreted using circular arc geometry will be shallower than the actual target.

The usefulness of redundant pole–dipole resistivity profiles is evident in the survey results described above; namely, no *a priori* knowledge of the target location is required and both position and depth along the traverse can be derived from the analysis. By demanding that several circular arc intersections, e.g. three or more, accumulate at an anomalous subsurface location before declaring it a suspected target, the redundancy of the survey data is used advantageously to enhance the validity of target interpretation.

Processing and analysis of the automatically recorded field data collected by the ERDAS equipment employs a more elaborate and powerful technique to significantly enhance the detected target signal-to-noise ratio. In particular, the method utilises a digital computer to sort out the recorded data and to apply the appropriate geometric factors to obtain many highly redundant pole–dipole resistivity profiles. The computer also accepts auxiliary information pertaining to the topographical elevations of the current electrode positions along the survey traverse for the purpose of compensating for distortions in surface potentials measured in irregular terrain. Finally, a target anomaly matched filtering technique is applied whereby computer-generated model resistivity profiles associated with many different assumed subsurface cavity target locations are compared with the experimental field data via a cross-correlation process.

Maximum correlation between the target anomaly and the anomaly signature predictions occurs when the model target is positioned to give the closest relative correspondence among all of the overlapping measured and

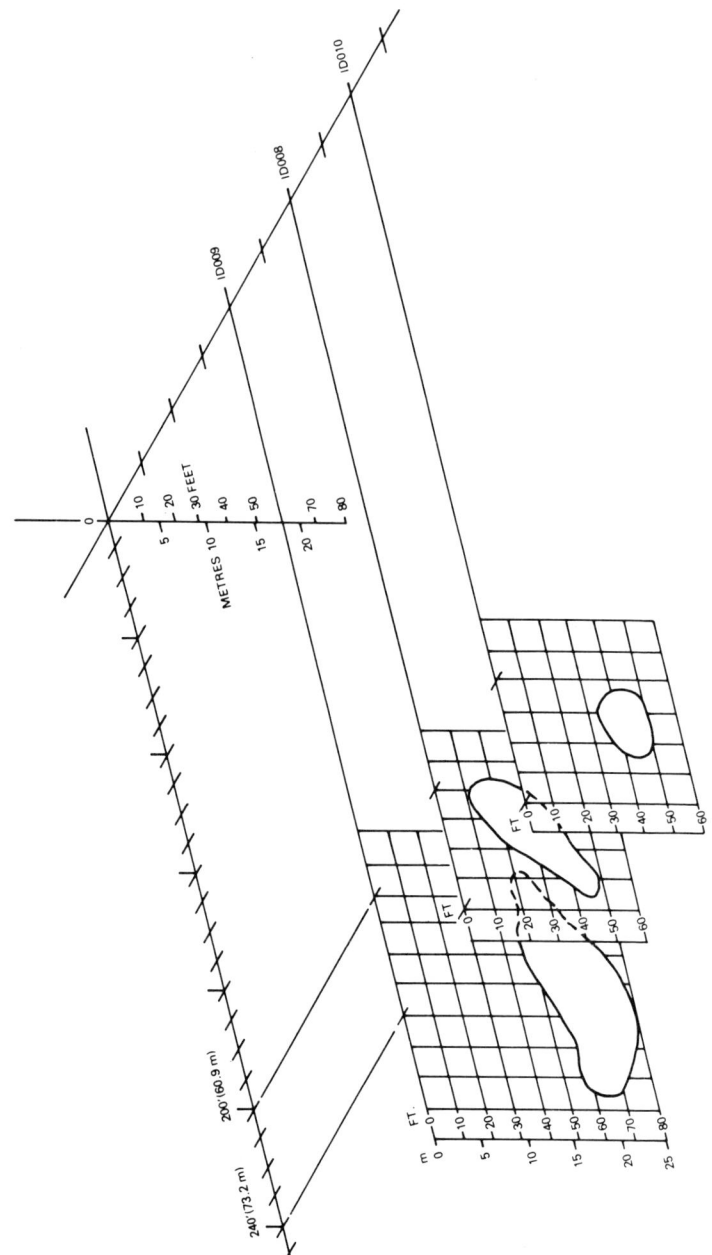

FIG. 16. Three-dimensional interpretation of resistivity survey over Medford Cave.[7]

predicted apparent resistivity profiles. Thus, the target model location which yields the maximum correlation with the measured data also indicates the most probable location of the suspected target. Further, the magnitude of correlation provides a confidence factor regarding the degree of correspondence between the assumed model target (ideal desired target response) and the suspected anomaly detected.

The results of this predictor–correlator process are carried out in a flat-halfspace geometric model in which all of the actual field terrain elevations and electrode coordinate positions have been conformally mapped by the Schwarz–Christoffel transformation. In this case, the warped coordinates of the transform domain permit a signature predictor model such as that described earlier by eqn (7) to be used in determining the surface electric fields and, hence, the predicted surface potentials that would be measured by means of the terrain-corrected potential electrode locations.

Next, since these predicted potentials do not change when mapped back to the actual terrain elevation conditions they are reduced directly to relative resistivity profiles for each current station by applying the same geometric factors applied to the experimental potentials recorded in the field. The resistivity profiles predicted for each current station, i.e. a forward and a reverse profile, and then numerically processed together with the experimental field data to yield the cross-correlation coefficient and, hence, a quantitative measure of similarity between the entire ensemble of resistivity profiles measured in the field and those derived for the particular model target position assumed in the analysis. By moving the assumed model target position to successive target resolution cells in the subsurface zone of interest and repeating the model versus experimental resistivity data-correlation process, one or more isolated positions of maximum correlation will be found, designating the most likely cavity target locations. Figure 17 illustrates the automatic-resistivity data-processing technique in diagrammatic form. In practice, this process has been implemented on a minicomputer having a 16K-word core memory and a 2·5M-word disk memory. Computational time using this modest capability to process a 500-m traverse line varies from a few hours to one or two days depending upon the number of vertices used in the Schwarz–Christoffel transformation, the depth of investigation and the selected size of the target resolution cells. The results of this analysis can be displayed in an image-like format by plotting the cross-correlation coefficients designated for each target position resolution cell in the form of a shade-of-grey scale or as contour lines representing an anomalous resistivity target under the survey traverse line.

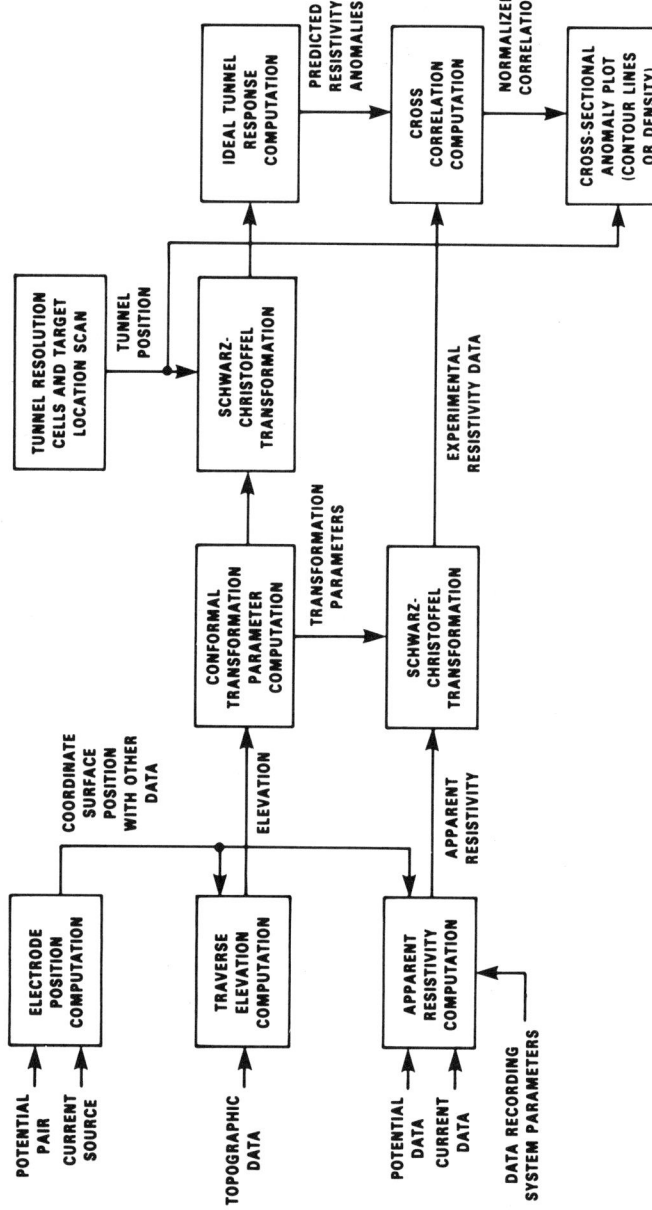

FIG. 17. Automatic earth resistivity data processing program for tunnel detection.

Figure 18 shows the results of an automatic resistivity field survey conducted over a 3-m diameter water-diversion tunnel located in granite rock (the tunnel was temporarily free of water). This tunnel was constructed through the sloping side of a mountain and, therefore, the opportunity was available to measure the target response at various target depths below the surface. The diamond symbols represent the surveyed traverse elevations at each current electrode location adjusted to tunnel axis depth. The shallowest survey was at 20 m as shown in Fig. 18(a) which indicates a very accurate detection and location of the tunnel (target axis coordinates are (0, 0) in depth and traverse horizontal position). For those traverses located further up slope as shown in Figs 18(b)–(d), the tunnel target is clearly detected at successively deeper depths without serious false alarm responses. However, it is noted that the resolution of the method is less at increased target depths; as would be expected from potential theory. The cross-correlation coefficient varies from 0·65 for the shallowest target to about 0·4 for the deepest target, showing that the relative detection confidence also degrades with target depth.

Errors in target position and depth as shown for the deepest tests are believed to be the result of non-uniform and possibly anisotropic resistivity conditions in the relatively thick overburden rock above the tunnel. Nevertheless, the fact that only one maximum correlation response is present in the displays implies that the matched filtering process is particularly effective in suppressing non-tunnel target responses in the processed results. Further evidence of the improved data-processing power of this method is provided by the fact that for the deepest targets shown in Fig. 18 the maximum perturbation in apparent resistivity, according to the approximate relationships given in eqn (6) is in the range of about $(\rho_a/\rho_0)_{max} = 0·08\%$ to $(\rho_a/\rho_0)_{max} = 0·19\%$, closely corresponding to the practical limits of accuracy of the field measurements and recorded data.

The automatic resistivity system has been used to survey several shallow abandoned coal mine workings in Kentucky, Indiana, Wyoming and Colorado.[10,11] These surveys have been generally successful in detecting both air-filled and water-filled sections of retreat-mined panels and coal mine entry ways in the depth range of 60–80 m. At the present stage of development, the data-processing program is not designed to include stratified geologic earth structure models or to accurately model large-scale cavity targets such as a room and pillar mine working. Nevertheless, the processed data has provided results which correspond very closely with the available mine maps and, in the case of surface evidence of mine

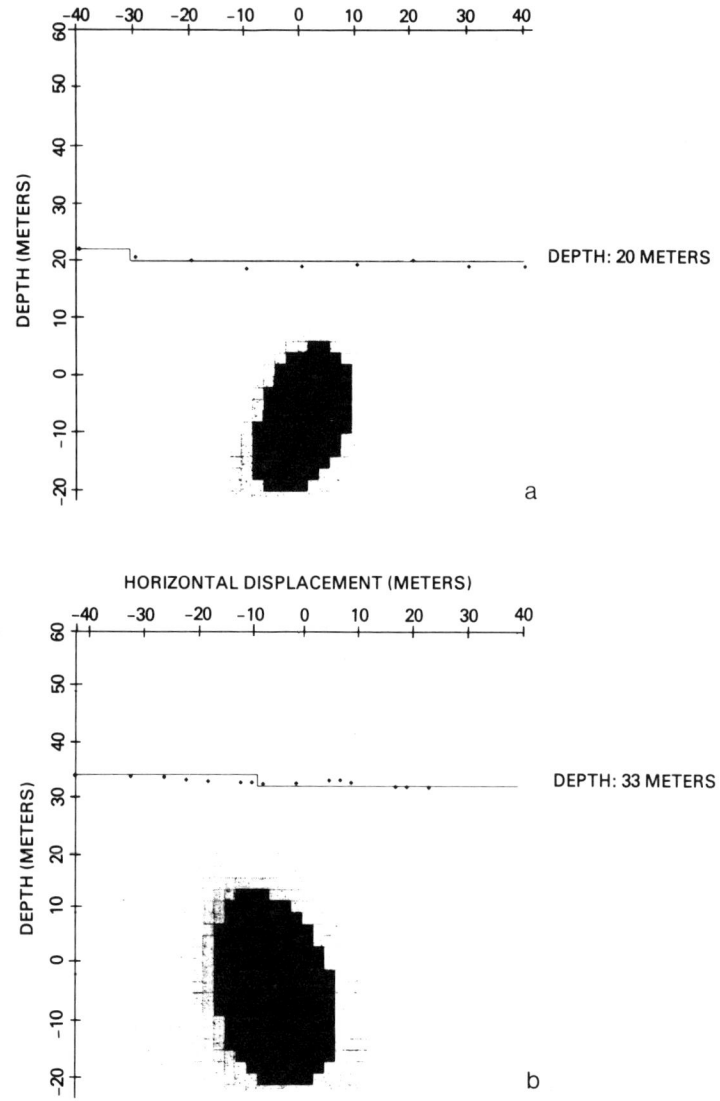

FIG. 18. Earth resistivity survey results over a water diversion tunnel near Basalt, Colorado (3-m diameter tunnel in granite; centre-line at 0, 0).[10]

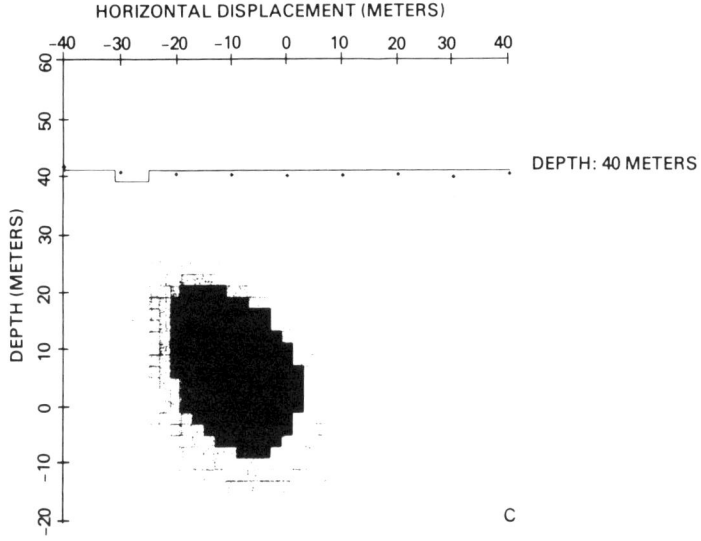

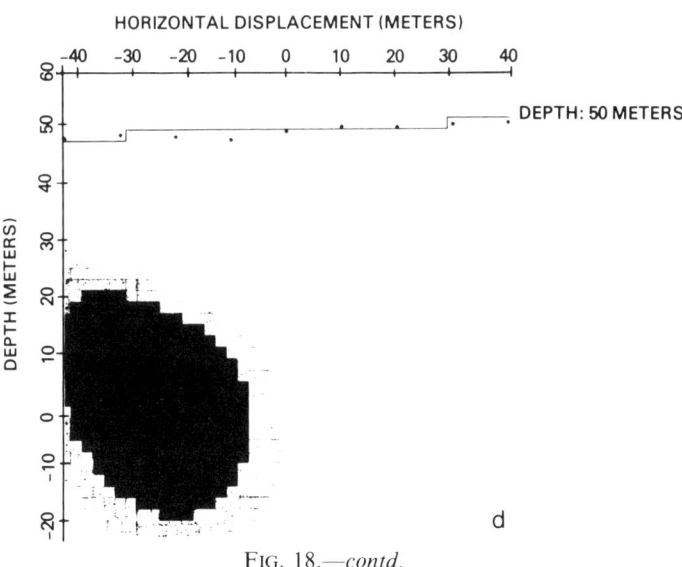

Fig. 18.—*contd.*

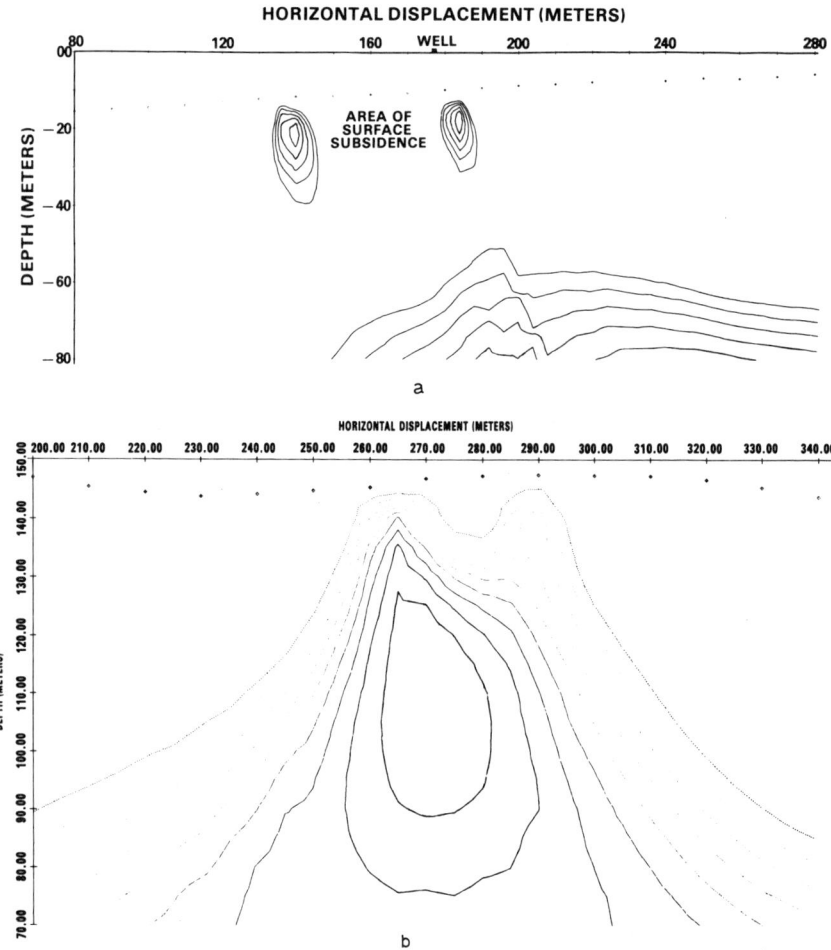

FIG. 19. High-resistivity anomalies detected in shallow coal mine workings. (a) Acme mine, Sheridan, Wyoming; (b) coal mine entry way, Kentucky (line no. 3).[10,11]

overburden subsidence, with the probable location of unmapped underground mine openings or rubble zones.

Figures 19 and 20 present examples of these coal mine survey results in which the data were processed for both high-resistivity and low-resistivity target conditions. The diamond symbols in both of these figures represent the surveyed traverse elevations of each current source electrode location relative to reference elevations at the starting ends of each traverse.

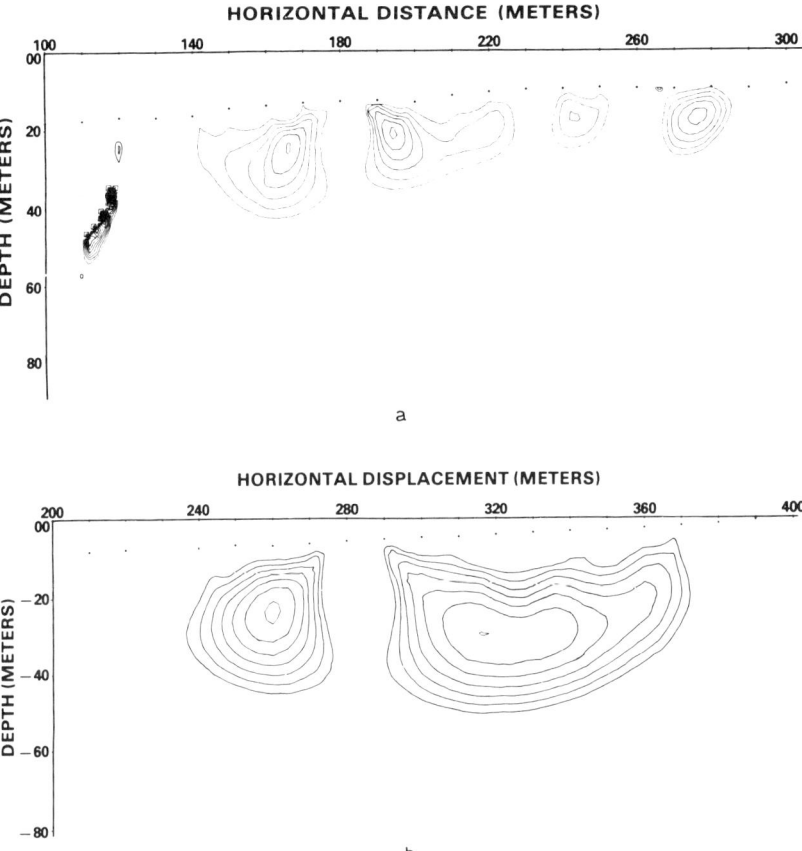

FIG. 20. Low-resistivity anomalies detected in shallow coal mine workings. (a) Monarch mine, Sheridan, Wyoming; (b) Acme mine, Sheridan, Wyoming.[10,11]

The high resistivity anomalies identified in Fig. 19(a) and (b) were determined to be deeper than indicated in the processed results; however, the anomaly positions along the survey traverses were in accurate agreement with information obtained from available maps of the mines. The results obtained from low-resistivity anomalies corresponded with known water-flooded zones of the mines or zones of roof collapse where water-saturated rubble is assumed to be present. As shown in Fig. 20(a), the low-resistivity targets agree reasonably well with the high-resistivity anomalies (such as that shown in Fig. 19(a)) and, hence, may be assumed to be associated with parts of the same coal seam workings. A limited amount of evidence from other surveys indicates that when the general resistivity profile of the geological medium increases with depth, the data-processing technique based upon a homogeneous medium shows cavity targets that are located shallower than their actual depths.

4. SURFACE SEISMIC METHODS

Surface seismic exploration techniques have been applied to cavity detection and mapping using concepts of compressional wave attenuation observed as shadows in deeper marker bed reflections,[23] high-resolution cavity reflections,[24] and the cavity walls as well as flexural vibrations of the roof overburden.[25] While these studies have been informative in establishing the practical potential of seismic exploration technology, the results have not proven reliable. High-resolution seismic reflection techniques continue to receive attention, however, with the objective of detecting targets such as shallow abandoned underground mine workings. In this application, sources consisting of small explosive charges in the range of one to five ounces have been used and various controlled-waveform vibrators and impulse sources are under investigation for mapping coal mined areas at depths to 500–600 ft.[26]

Shear waves and Rayleigh waves have also been investigated in connection with coal mine cavities. Refraction methods using shear waves along continuous higher velocity layers underlying the mined area were found to yield shadows of certain mine cavity areas in the observed refraction profiles.[27] Amplitude diffraction and travel-time delay effects caused by obstacles located in the path of Rayleigh wave propagation have proven to be reliable in detecting and locating near surface cavities.[28]

From a practical view, conventional seismic exploration methods have not been successful in cavity detection because of the marginal resolution attainable with respect to the typical sizes of cavity targets. With the exception of the Rayleigh wave-diffraction concept mentioned above and similar diffraction methods using body waves generated and detected in boreholes, no other seismic techniques have been specialised for detecting cavity targets. The application of surface seismic techniques in a search mode of operation must therefore await the adaptation of the emerging high-resolution reflection methods to the difficult problems associated with relatively small cavity targets often located in weathered or disturbed geological media.

When the cavity is shallow such as in the case of covered and abandoned vertical mine entrance shafts, Rayleigh wave-diffraction techniques have been used as a successful means of detection and position location. Dresen and Hsieh[28] conducted extensive three-dimensional scale model studies of through-transmission body waves and Rayleigh waves using a point source and a single receiver located on the surface. Experiments simulating open and buried vertical mine shafts having a relative diameter of approximately 1·5 wavelengths at the Rayleigh wave-signal frequency exhibit distinctive amplitude diffraction patterns when the shaft target is located between the source and receiver. By scanning the receiver in an angular fan pattern at constant radius from the source, the spatial distribution of the received Rayleigh wave amplitude exhibits a W-shaped curve when plotted versus scan angle. Figure 21 shows the test measurement layout used in the model studies and in similar field surveys.

Figure 22(a) shows test results obtained using a two-layered model in which refracted and reflected body waves arriving at the detector via the subsurface interface as well as Rayleigh waves which are not necessarily dependent upon the layer interface are diffracted by the shaft. The normalised amplitude responses in this figure show the well-defined W-shaped spatial pattern measured over a $\pm 20°$ scan. The minima in response are essentially the same for all of the observed wave types, and for the relative shaft size used in this case, occur at angles slightly less than the shadow angle formed by the shaft target as illuminated by the source. The ratio of the central maximum to the minimum amplitudes in the W-shaped curves is about two for Rayleigh waves whereas the reflected and refracted waves have amplitude ratios of about 1·4 or less.

Figure 22(b) shows the relative travel time delays versus angle for Rayleigh waves and refracted and reflected body waves in the model studies. The maximum time delay occurs for Rayleigh waves when the detector is directly aligned with the source and the shaft target. For the model example shown, the peak values of travel time delay are approximately 3% of the total propagation time.

A practical field test reported by Dresen and Hsieh[28] demonstrated the Rayleigh wave method of shaft detection at two different mine locations. Figure 23 shows the diffraction amplitude profiles and the travel time delay anomalies for the two mine shaft targets. The W-shape amplitude diffraction anomalies are evident in both cases as are the travel time delay effects. In the case of the known cross-section target at the Friederika shaft, an equivalent cylindrical shaft size of 2·8-m diameter was interpreted from two measurement profiles. The mine shafts at both field sites were

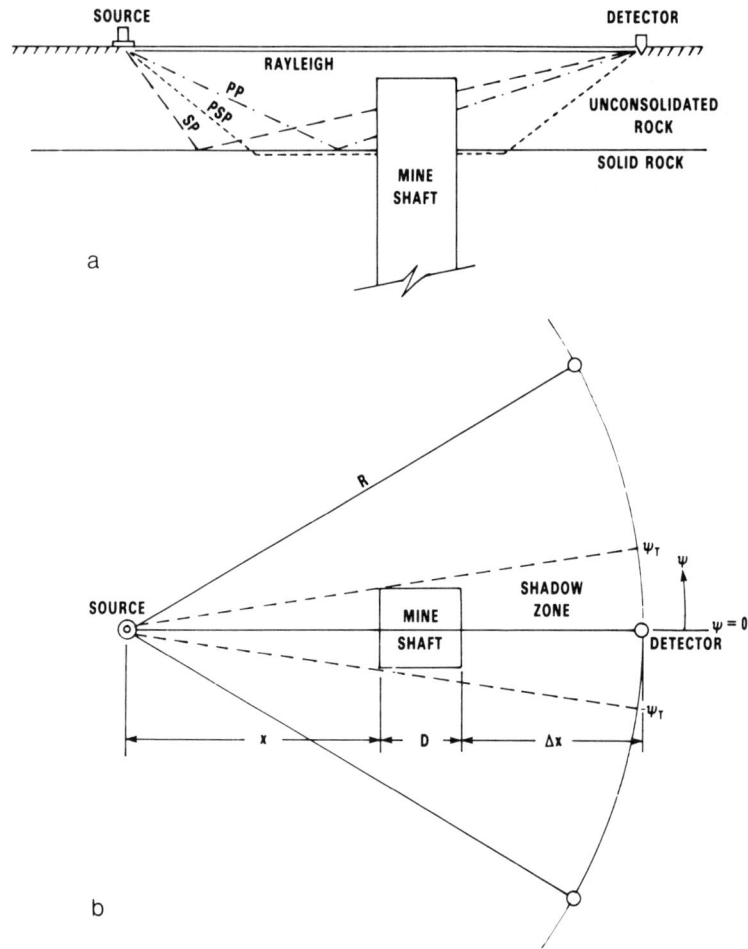

FIG. 21. Application concepts for Rayleigh wave seismic detection of shallow abandoned mine shafts. (a) Seismic wave propagation paths; (b) scan pattern layout on surface.[28]

accurately located using the Rayleigh wave-diffraction measurement technique.

This method of cavity detection using Rayleigh wave diffraction is an outgrowth of earlier work by Dresen and others in which the source and detector were located in vertical boreholes as a means of detecting horizontal shafts or cavities via diffraction effects on body waves. By eliminating the need for drill holes, the Rayleigh wave technique provides a

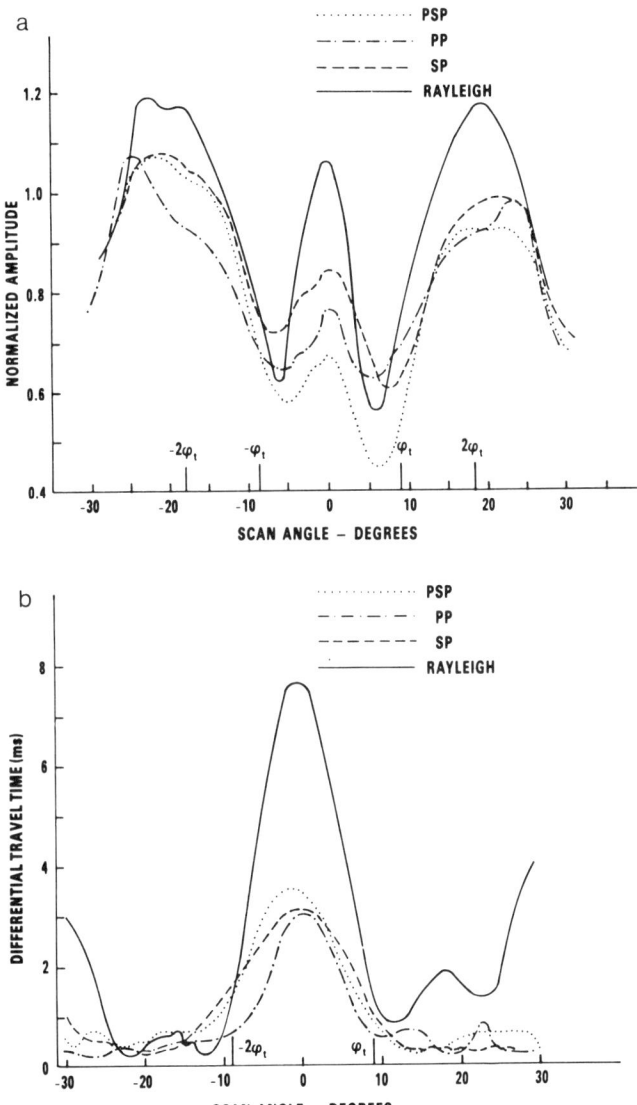

FIG. 22. Comparison of diffraction wave propagation anomalies for reflected, refracted, and Rayleigh waves (shaft is open to surface). (a) Diffracted wave amplitude versus scan angle; (b) diffracted wave travel time versus scan angle.[28]

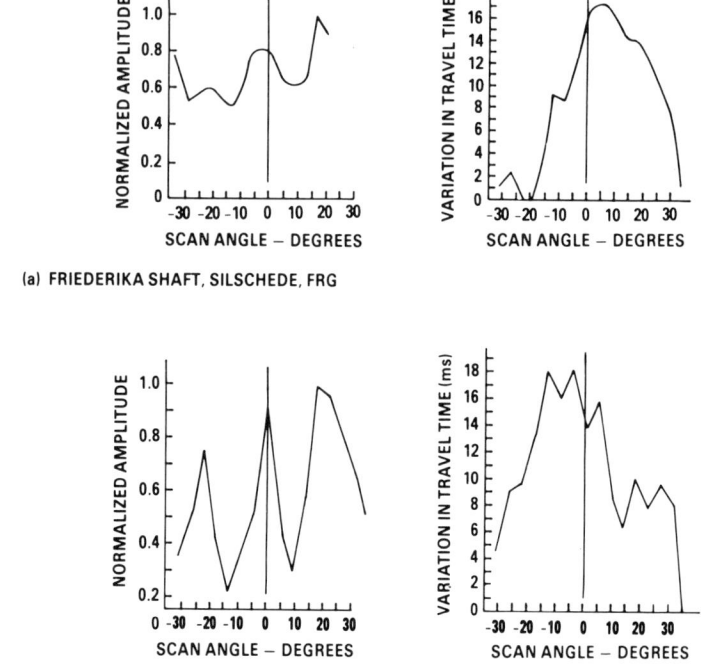

FIG. 23. Field test results of Rayleigh wave detection of shallow abandoned mine shafts.[28]

cost-effective approach to the search for and mapping of relatively shallow cavity targets.

5. SURFACE GRAVITY METHODS

Precision surface gravity surveys have been applied to the problem of cavity and tunnel detection in several ways. Colley[29] conducted one of the earliest detailed surveys and interpretations of limestone solution caves using gravity methods normally associated with petroleum exploration. Arzi[30] conducted a highly detailed microgravity survey of a nuclear power plant site to determine soundness of the near-surface bedrock. The work of Fajklewicz[31-33] however, is particularly extensive in regard to the detection of caverns and shallow abandoned mine workings. Dresen et *al.*[34,35] have reported improved instrumentation techniques for applying vertical gravity gradient measurements for detection of subsurface cavities. Butler[36] has investigated the application of microgravity and gravity gradient techniques to problems of shallow geotechnical investigations, including the detection and mapping of limestone solution cavities. New developments in high-precision gravity meters within the past several years have allowed steady advances in the accuracy and speed of gravity field measurements. In this regard, advanced equipment and the use of multiple readings and precision data reduction can provide practical gravity measurements having an accuracy of about $\pm 5\,\mu$gal corresponding to changes in the earth's total gravitational field (about 980 gals) of about five parts per billion.

Cavity detection by means of gravity measurements requires the sensing of localised negative gravity anomalies in a background of larger scale regional gravity profiles or contours. Air-filled cavities offer the largest anomaly condition because of the complete absence of dense material in the target. In this regard, the density anomaly of a cavity is expressed in terms of the density contrast with the surrounding geological material; that is

$\Delta \rho = \rho_{\text{cavity}} - \rho_{\text{medium}}$

$\simeq -2 \cdot 5 \times 10^3 \, \text{kg/m}^3$ for air-filled cavities

$\simeq -1 \cdot 5 \times 10^3 \, \text{kg/m}^3$ for water-filled cavities

$\simeq -1 \cdot 0 \times 10^3 \, \text{kg/m}^3$ for rock rubble or mud-filled cavities (8)

In comparison with air voids, water-filled cavities offer a gravity anomaly effect which is only about 60% of that of the same cavity containing air; a rubble- or mud-filled cavity anomaly is about 40% that of air.

Despite the sensitivity of modern gravity instruments, cavity targets of practical size and depth are often near the threshold of detection because of their modest perturbations of the earth's gravitational field. Variations in subsurface geology and surface topographical relief can introduce noise in the measurements which may exceed the gravity anomaly effects of cavity targets when their spatial dimensions and proximity to the gravity meter are comparable with the target size and depth parameters.

More elaborate field survey procedures consisting of additional spatially-distributed readings can aid in resolving the target in the presence of noise, particularly where the approximate size and depth of the target being sought can be used as a guide in specifying the survey grid. Multiple traverses with closely spaced measurement stations and relative elevations measured to within 0·01 ft will result in a useful spatial redundancy and data accuracy that can be analysed to separate the anomaly caused by the target from the geological and topographical noise. In addition to the anomalous gravity effect, the vertical gradient of gravity can be determined from readings at two or more elevations at the same location. The vertical gradient of gravity can be used for cavity detection with the advantage that small shallow targets can be separated from larger scale anomalies. The anomalous gravity effects of typical cavity targets can be computed directly for idealised geometrical shapes in a homogeneous medium. For an infinite cylindrical tunnel oriented horizontally in a flat halfspace, the gravity effect observed along a transverse survey path (x-axis) is

$$\Delta g(x)_{\text{cyl}} = \frac{200\pi G \Delta\rho R^2 D}{D^2 + x^2} (\text{gal}) \tag{9}$$

where G = gravitational constant ($6\cdot 67 \times 10^{-11}\,\text{NM}^2/\text{kg}^2$); 1 gal = 10^{-2} N/kg; $\Delta\rho = \rho_c - \rho_0$ = density difference between that of the cavity and the host medium (kg/m^3); R = cylinder radius (m); D = depth from instrument datum to cylinder axis (m); and x = instrument offset distance from cylinder axis (m). Correspondingly, for a spherical cavity, the gravity effect is

$$\Delta g(x)_{\text{sph}} = \frac{400\pi G \Delta\rho R^3 D}{3(D^2 + x^2)^{3/2}} (\text{gal}) \tag{10}$$

Figure 24 presents normalised gravity effect profiles for air-filled and water-filled tunnels in a typical geological medium of density $2\cdot 5 \times 10^3$ kg/m^3. For example, a 1·5-m radius air-filled tunnel at a depth of 10 m has a maximum gravity anomaly of $2\cdot 25 \times (-10\cdot 5) = -23\cdot 6\,\mu$gal. The same cavity filled with water has a maximum gravity effect of $-14\cdot 2\,\mu$gal. Figure 25 presents similar gravity effect profiles for air-filled and water-filled spherical cavities. These curves show that a 2-m radius spherical cavity at a depth of 5 m has a maximum gravity effect of $-22\cdot 4\,\mu$gal when filled with air and $-13\cdot 4\,\mu$gal when filled with water.

A practical problem often encountered in seeking such small gravity anomalies is that larger-scale variations in deeper subsurface structure will produce regional trends in the measured gravity effect which can obscure the signatures of smaller anomalies of interest. This form of target masking can be reduced by measuring the approximate vertical gravity gradient by taking two readings at slightly different instrument heights at each survey station.

The vertical gravity gradient is derived approximately from two

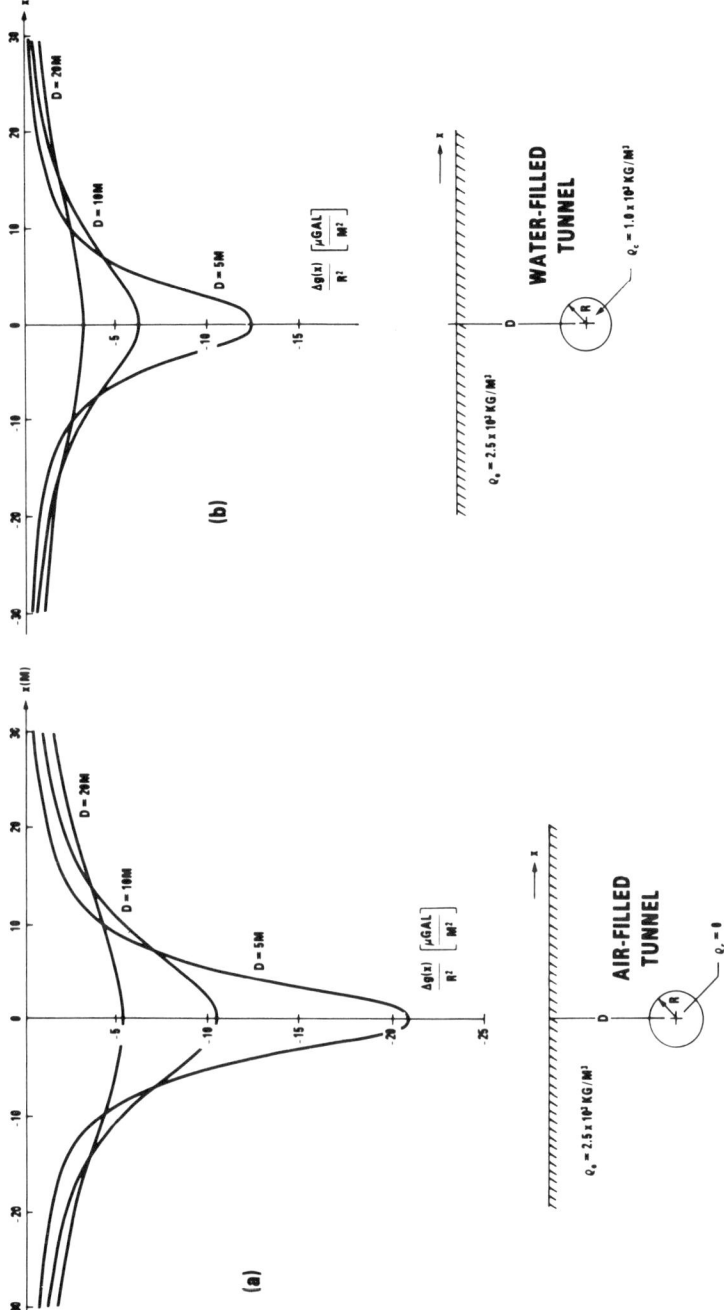

FIG. 24. Gravity effect profiles for air- and water-filled cylindrical tunnels (normalised to cylinder radius squared). Depth from instrument datum to cylinder axis = D (m).

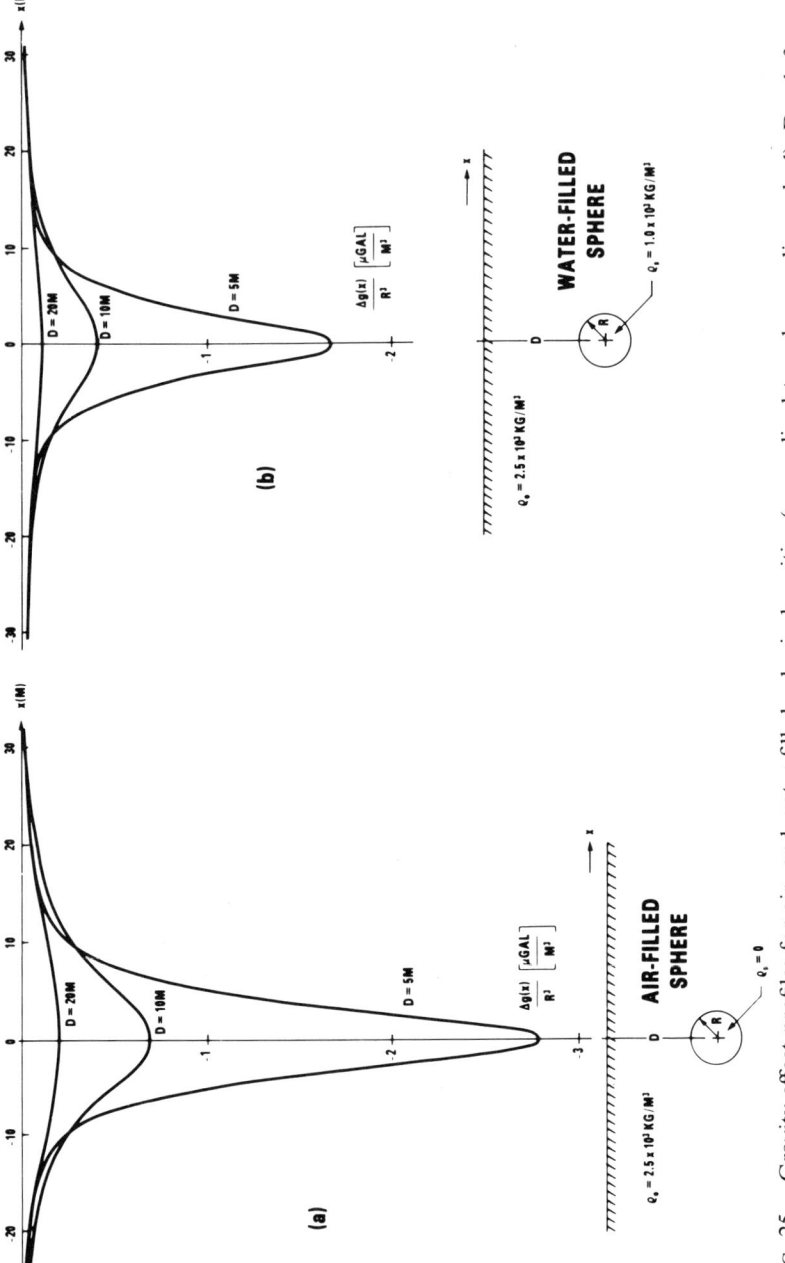

FIG. 25. Gravity effect profiles for air- and water-filled spherical cavities (normalised to sphere radius cubed). Depth from instrument datum to centre of sphere = D(m).

instrument readings taken at an elevation difference of $z_2 - z_1 = \Delta h$, as the differential quotient

$$\frac{\partial \Delta g(x)}{\partial z} \simeq \frac{\Delta g2 - \Delta g1}{\Delta h} = g_{zz}(x) \quad (11)$$

For cylindrical and spherical targets, the vertical gravity gradient profiles are

$$g_{zz}(x)_{\text{cyl}} \simeq \frac{\Delta g(x)_{\text{cyl}}}{\Delta h}\left[1 - \frac{1 + \dfrac{\Delta h}{D}}{1 + \dfrac{2D\Delta h}{D^2 + x^2}}\right] \quad (12)$$

and

$$g_{zz}(x)_{\text{sph}} \simeq \frac{\Delta g(x)_{\text{sph}}}{\Delta h}\left\{1 - \frac{1 + \dfrac{\Delta h}{D}}{\left(1 + \dfrac{2D\Delta h}{D^2 + x^2}\right)^{3/2}}\right\} \quad (13)$$

Normalised vertical gravity-gradient profiles for cylinder and spherical cavities are shown in Figs 26 and 27. Vertical gravity gradient is often expressed in units of Eötvös (1 E = 0·1 μgal/m). For example, the gravity-gradient curves of Fig. 26 show that a 1·5-m diameter air-filled tunnel at a depth of 5 m has a maximum gradient anomaly of $2·25 \times (-2·97) = -6·7\,\mu\text{gal/m}$ (-67E) for $\Delta h = 1$ m. The gravity effect readings which yield this result are $-47·1\,\mu\text{gal}$ and $-39·3\,\mu\text{gal}$ (from eqn (9)) at heights of 5 m and 6 m, respectively; values well above the typical instrument threshold of $\pm 5\,\mu\text{gal}$. Figure 27 indicates that a spherical air cavity of 2-m radius at a depth of 5 m has a maximum gradient anomaly of $8 \times (-0·78) = -6·2\,\mu\text{gal/m}$ (-62E) for $\Delta h = 1$ m as determined from gravity effect readings of $-22·4\,\mu\text{gal}$ and $-15·5\,\mu\text{gal}$ (from eqn (10)) at instrument heights of 5 m and 6 m, respectively. Figure 28 presents normalised curves of the maximum negative gravity effects and vertical gravity-gradient anomalies for cylindrical and spherical cavities versus depth.

Dresen et al.[34] show that, for a Gaussian error distribution in the gravity readings, the error in the differential quotient is

$$F(g_{zz}) = \sqrt{\frac{2}{n}}\frac{F(\Delta g)}{\Delta h} \quad (14)$$

where $F(\Delta g)$ = error in gravity instrument readings; n = number of independent gravity readings at each elevation; Δh = elevation difference in

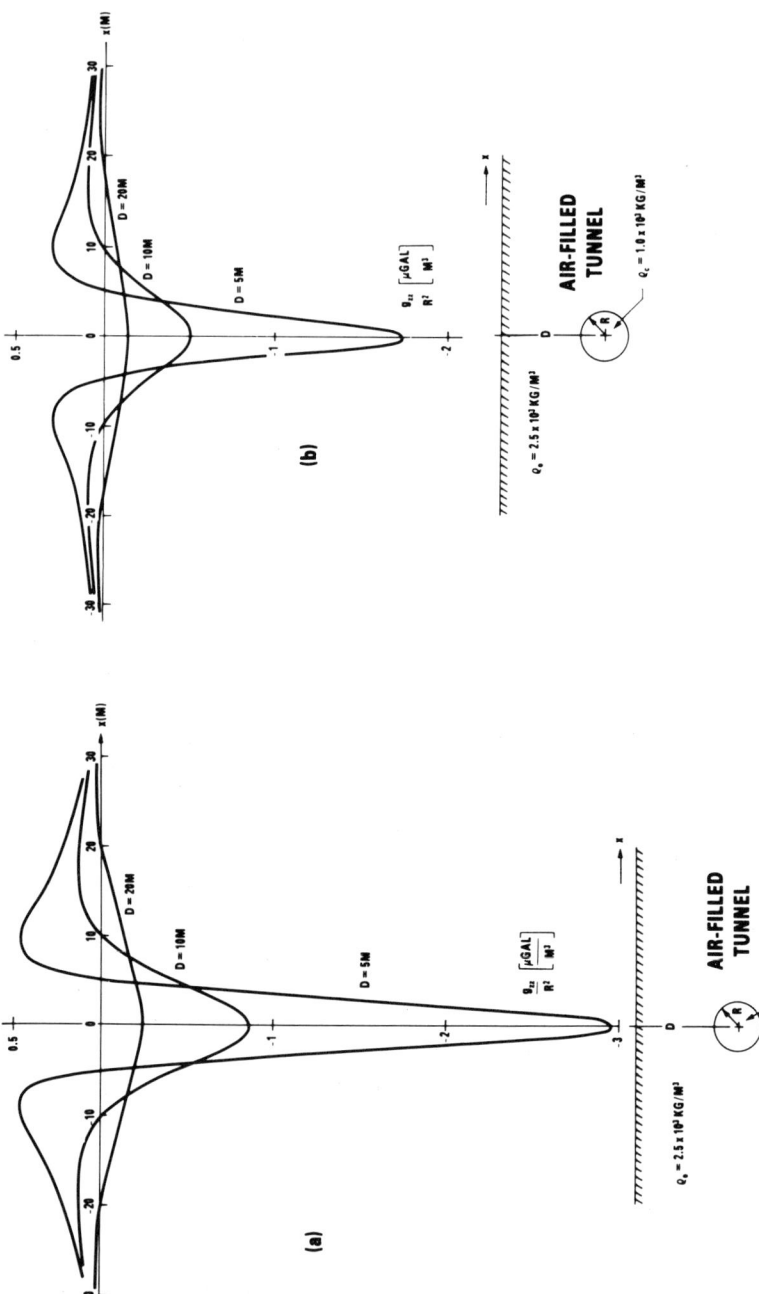

FIG. 26. Vertical gravity gradient (differential quotient) for air- and water-filled tunnels (normalised to cylinder radius squared). Instrument height difference $\Delta h = 1$ m. Depth from instrument datum to axis of tunnel $= D$ (m).

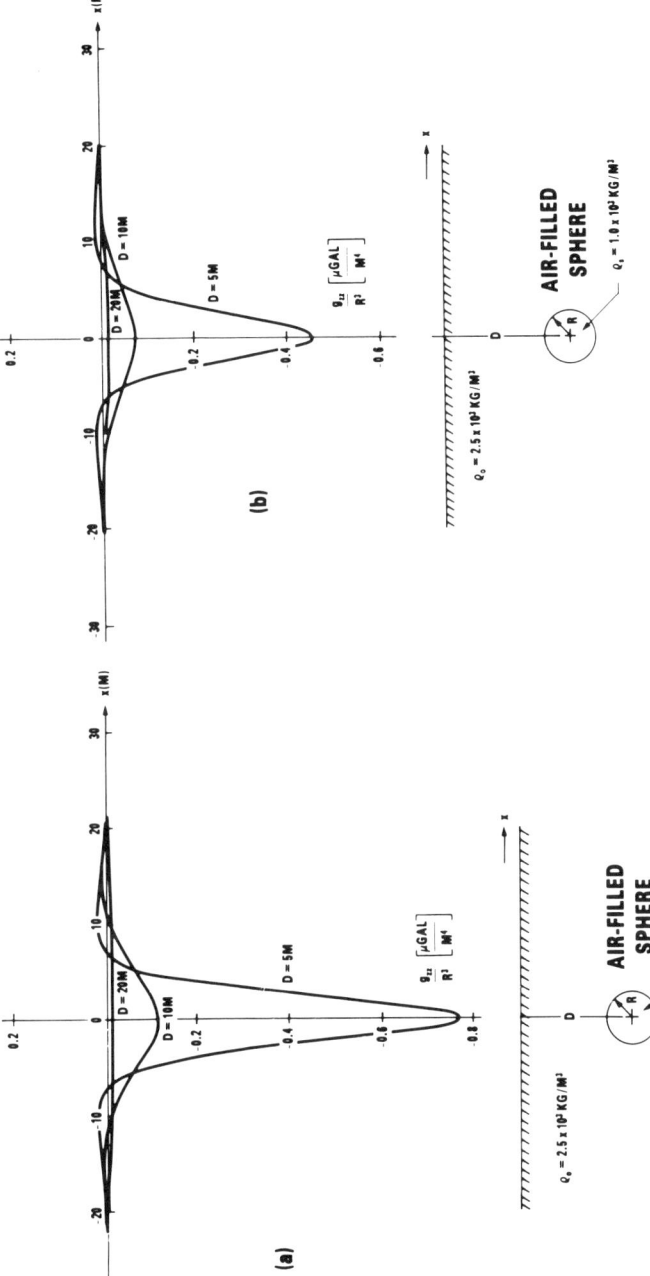

FIG. 27. Vertical gravity gradient (differential quotient) for air- and water-filled spherical cavities (normalised to sphere radius cubed). Instrument height difference $\Delta H = 1$ m. Depth from instrument datum to centre of sphere $= D$(m).

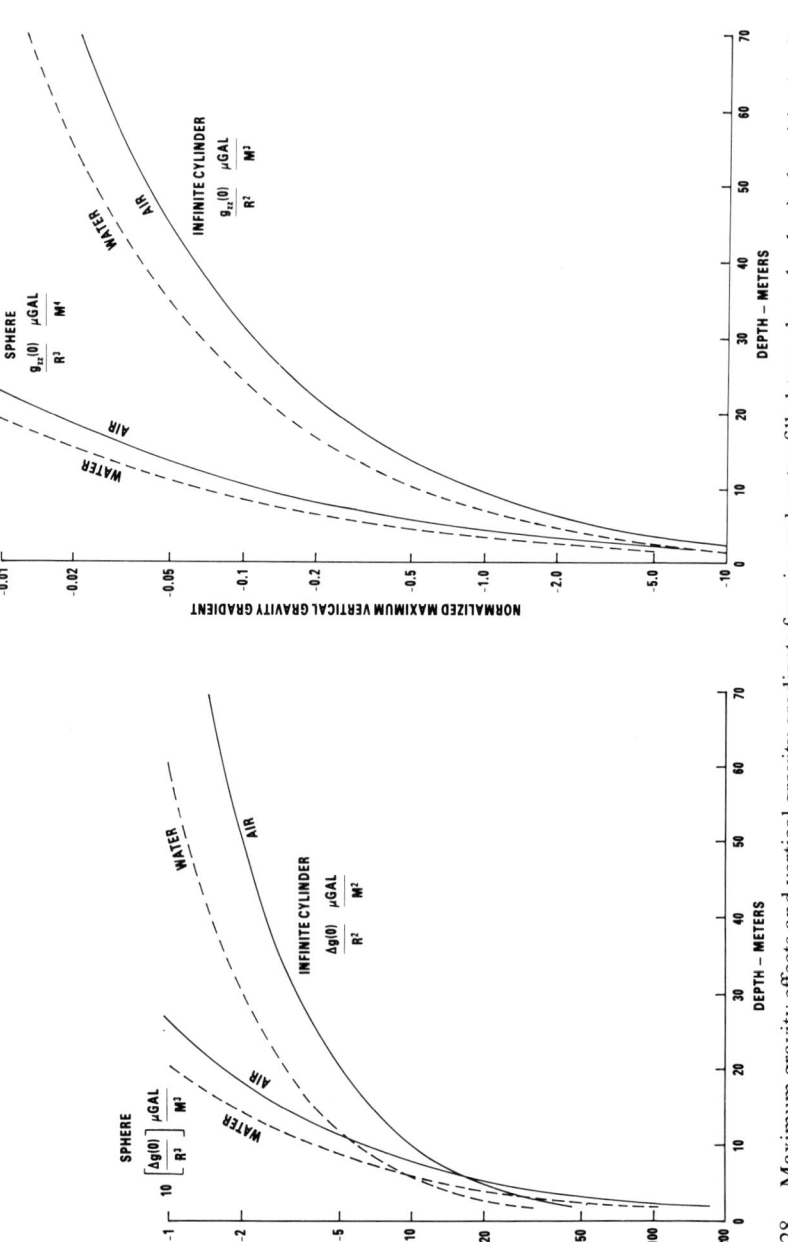

FIG. 28. Maximum gravity effects and vertical gravity gradients for air- and water-filled tunnels and spherical cavities (cavity radius (m); $\Delta h = 1$ m).

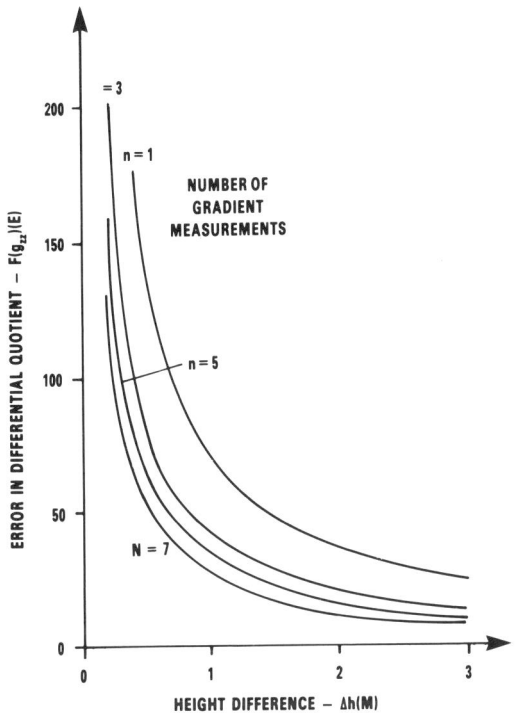

FIG. 29. Error range of gravity-gradient (differential quotient) measurements versus instrument height difference and number of independent readings.[34,35]

instrument positions. Figure 29 shows the error in differential quotient versus instrument height differential for various numbers of repeated gravity meter readings.

From the relationship in eqn (14) and Fig. 29, the error in differential quotient is inversely proportional to the instrument elevation difference, indicating that a larger instrument height separation will reduce the error in the gravity gradient. Additional readings at each instrument height will also reduce the error in gravity gradient and, thus, relatively inaccurate results at a small value of Δh can be compensated by repeating the readings a sufficient number of times. For example, four pairs of instrument readings, as required to provide four derived values of differential quotient for a given difference of height, Δh, will provide the same composite accuracy as one pair of readings at an instrument difference of height of $2\Delta h$. However, the increase in time required to obtain the additional readings is the penalty paid for using the smaller instrument height difference. Dresen et al.[34]

FIG. 30. Vertical gravity-gradient instrumentation for precision cavity detection surveys.[34]

indicate that a measurement time of 3–5 min for a single complete measurement of differential quotient is possible using the specialised gravity-gradient equipment shown in Fig. 30.

This arrangement incorporates a Lacost and Romberg model D-34 gravimeter and an instrument support tower capable of providing a difference of height of $\Delta h = 1\cdot 45$ m. With this arrangement, Dresen *et al.*[34] claim an instrument accuracy of 5 μgal in the presence of typical environmental vibrations on the support tower. As a reasonable compromise between measurement accuracy in vertical gravity gradient and time and effort in the field, they[34] normally use three to five measurements at each instrument height, corresponding to a total time of 20–30 min per gravity-gradient station.

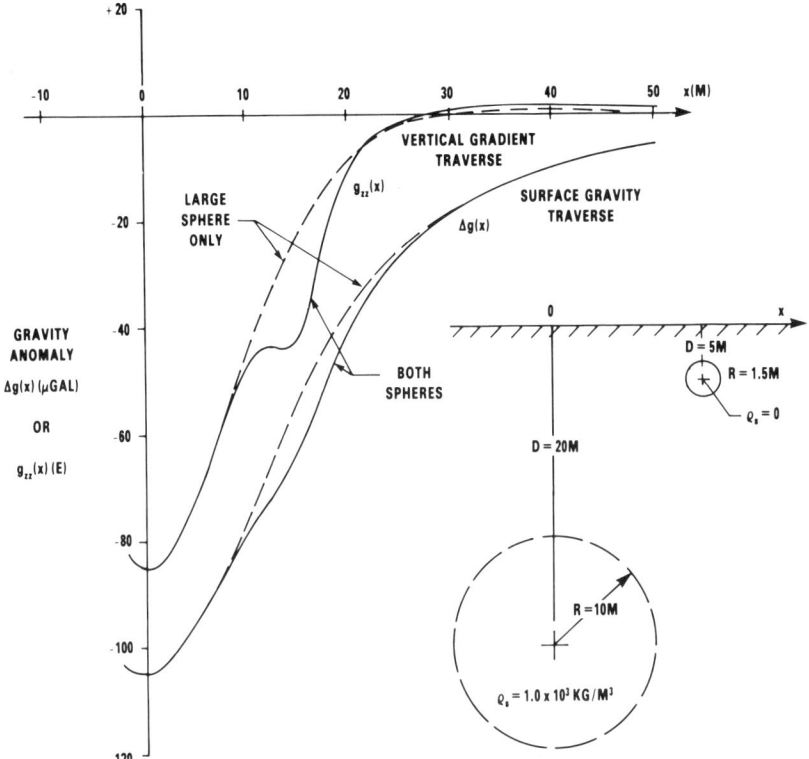

FIG. 31. Calculated gravity anomalies illustrating the detection of a small air-filled cavity in the presence of a larger water-filled cavity by using vertical gravity-gradient measurements ($\Delta h = 2$ m).

A simple example graphically demonstrates the advantage of vertical gravity-gradient measurements in Fig. 31 in which a small air cavity is located at an intermediate depth below the surface and above a larger and deeper low-density gravity anomaly. The masking effect of the large sphere is evident in the anomalous gravity effect in which the combined response of both spheres deviates only 17% from that of the large sphere alone. In contrast, the combined vertical gravity-gradient response exceeds the local gradient anomaly of the large sphere by 90%.

Fajklewicz[31] has applied the method of vertical gravity gradient in a variety of mining and engineering problems, particularly in connection with the detection of shallow abandoned mine workings and caves. In this work he employed an instrument support tower on which precision gravity

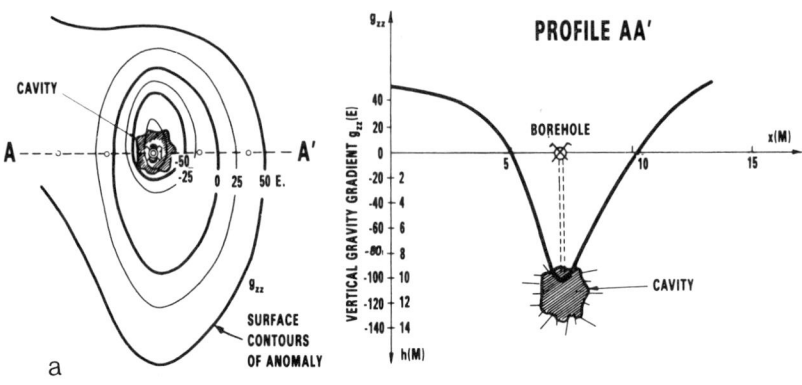

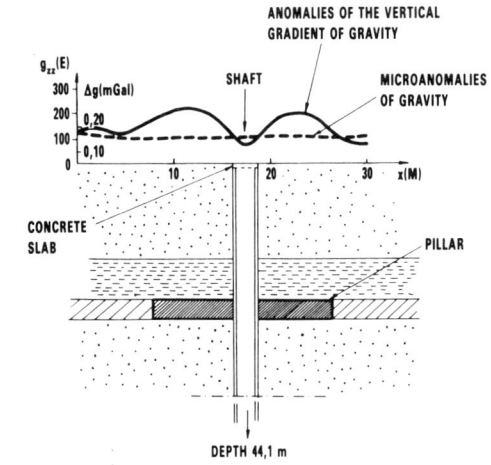

Fig. 32. Examples of gravity-gradient field survey results related to shallow mine cavities. (a) Post-exploration mining cavity; (b) abandoned mine entry shaft and subsurface safety pillar.[31]

gradient readings could be made at typical height differences of up to $\Delta h = 3$ m. With such a large instrument spacing, the differential quotient is only an approximation to the true vertical gravity gradient. Therefore, the method has been referred to as the 'gravity tower vertical gradient' method. The method is sensitive to shallow gravity features, and has been quite successful in detecting and mapping shallow cavities and old mining galleries at depths to about 15 m.

Figures 32(a) and (b) shows typical examples of differential quotient vertical gradient survey data and the subsurface cavity targets that caused the anomalies. These results were reportedly made in urban environments where a variety of cultural structures could introduce local gravity anomalies. However, because these anomalies are generally on, or very near, the surface and are located off the vertical axis of the instrument station, the differential quotient tends to remove those gravity effects that are essentially common to the instrument readings from which the vertical gravity gradient is derived.

6. BOREHOLE DETECTION METHODS

The foregoing surface geophysical survey techniques provide a means for searching for various cave and tunnel targets. Anomalies which are interpreted as cavities are normally checked by drilling. However, the drill may miss the target and the resulting drill holes can then provide a means for subsurface observations of the suspected target location. Hole-to-hole and hole-to-surface geophysical measurements capable of examining the intervening geological conditions between various source and detector locations offer the potential advantages of greatly improving the efficiency of the test boring programme by locating the targets more accurately via underground access and, where necessary, extending the cavity detection capability beyond the performance limitations of surface methods.

Four different borehole geophysical methods have shown promise in detecting caves and tunnels. These methods are electrical resistivity, electromagnetics, seismics and gravimetric techniques. In some cases, the methods only require one borehole for their operation or operate in a transmission mode between one hole and the surface. Other methods operate in a hole-to-hole transmission mode in which borehole scanning motions of either the source, the detector, or both are employed for the purpose of delineating the target anomaly. Further, the various methods are generally complementary in that the use of two or more different physical sensor responses offer improved confidence in verifying the presence of a cavity target prior to drilling additional test holes.

6.1. Borehole Resistivity Techniques

Borehole resistivity techniques can be subdivided into hole-to-hole, hole-to-surface, surface-to-hole and single-hole measurements. Additionally, the source and receiver electrode configurations employed in these different modes of application may consist of either fixed or scanned pole or bipole arrays as well as arrays designed to provide a focused electrical response.

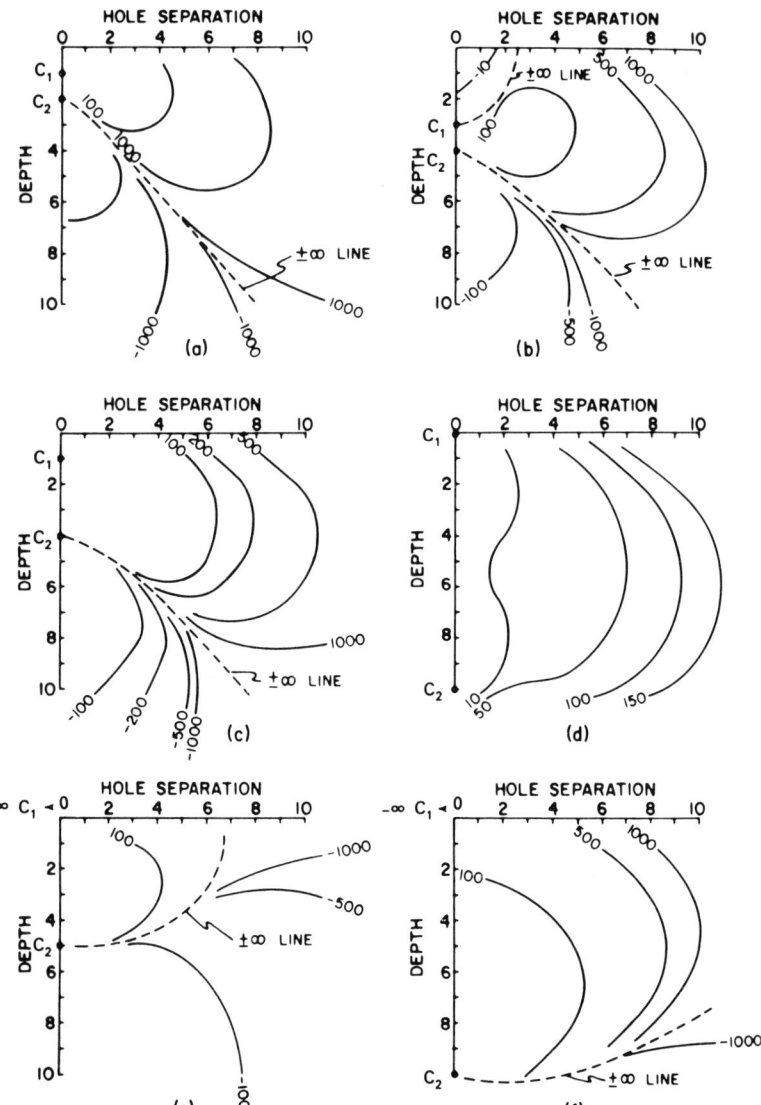

FIG. 33. Hole-to-hole resistivity surveys with current electrodes in one hole and receiver bipole in another hole. Curves show geometric factors required to convert the ratio of voltage to current to apparent resistivity. C_1 and C_2 are the current source electrode positions. The receiver bipole pair has a separation distance equal to the diameter of the spherical cavity target. The hole separation and depth scales are normalised to the cavity diameter. On the curves, depth is the depth to the centre of the receiver bipole. The dashed curve marked $\pm\infty$ results from a singularity corresponding to a zero potential line.[22]

Hole-to-surface measurements have been evaluated by several investigators[37,38] with reference to mineral exploration applications. More generalised studies of borehole resistivity techniques have been carried out by Daniels[22,39,40] using three-dimensional numerical modelling of spherical and lens-like target geometries and by Brieden et al.[41,42] and Militzer et al.[43] using analytical solutions for two-dimensional line electrodes and tunnel targets. Together, these studies cover 13 different electrode array and target combinations.

Daniels[22] points out in several cases and Militzer et al.[43] also mention the fact that, when both the source and detector electrodes are located below surface, the apparent resistivity may have a singularity (caused by the presence of a zero equipotential surface in the zone of potential measurements) depending upon the receiver array location relative to the source. However, in most cases the geometric conditions can be selected to avoid such singularities which could easily obscure the presence of a cavity target. Figure 33 shows six master curves depicting the value of the geometric factor as derived by Daniels[22] for six selected source electrode orientations in a homogeneous halfspace. The singularity conditions imposed by the electrode geometry are indicated by the dashed $\pm\infty$ lines, showing the conditions of receiver bipole depth range which should be avoided for a given hole separation.

In the presence of a geological target anomaly, the perturbations in apparent resistivity are essentially unaffected when the source-position-determined geometric singularity is greater than about twice the target depth for the hole separation used in the tests. Figure 34 shows this condition for a conductive target contrast for three downhole bipole source array positions and a hole separation of four receiver bipole spacings. In these examples, Fig. 34(a) shows the target anomaly to be masked when the source array and the target are at about the same depth whereas for the deeper geometric singularity shown in Fig. 34(b) the target is clearly evident. For the hole spacing used in these examples, the amplitudes of the resistivity anomalies range from about 0.5% to about 1%, depending upon the source electrode array and target depth. The peak values of the anomalies occur approximately at the depths of the corresponding targets. However, these responses do not provide an indication of the lateral position of the target with respect to the boreholes.

Geometric singularities can be avoided by scanning the downhole source and receiver arrays at the same depth below surface. With this arrangement, the target anomaly will always be observed without interference, independent of its depth. Figure 35 shows the resistivity

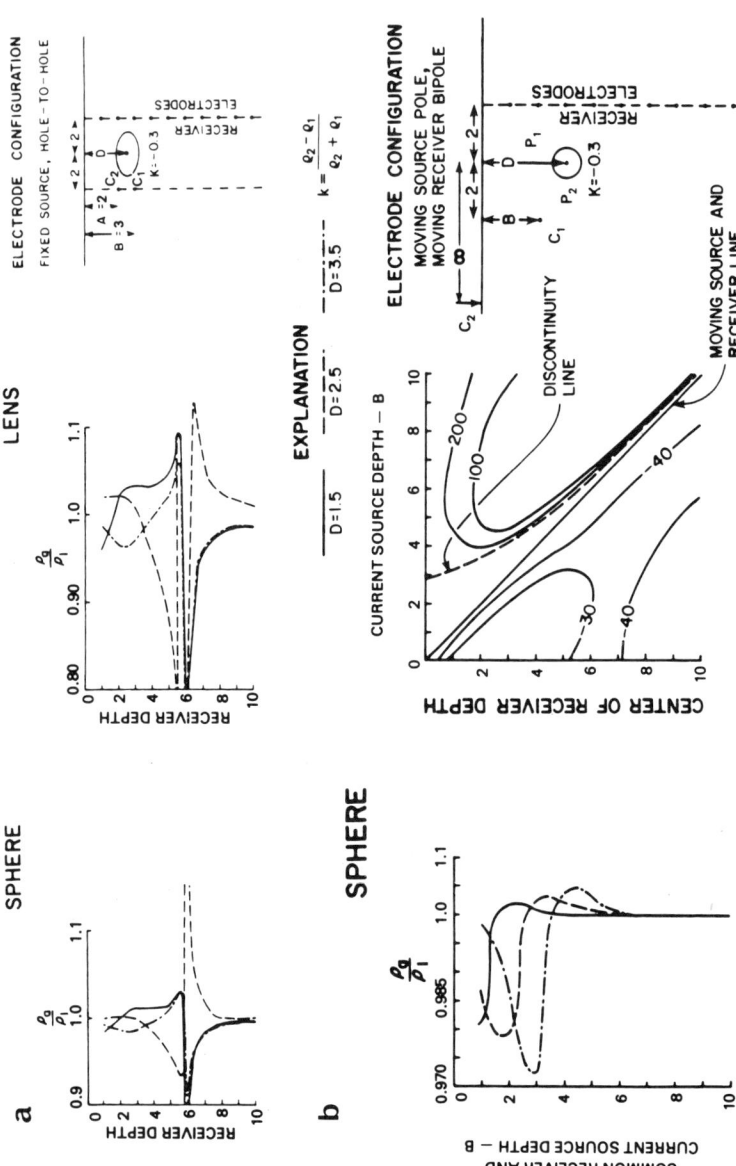

FIG. 34. Normalised apparent resistivity profiles for hole-to-hole measurements. Depth of electrodes is as defined in Fig. 33. (a) Stationary bipole current source in one hole and moving bipole receiver in adjacent hole (sphere diameter = 1 depth unit; lens: 2 units wide; 4 units long; 0·4 units thick; $K = (\rho_2 - \rho_1)/(\rho_2 + \rho_1)$). (b) Moving pole current source (C_1) in one hole and moving bipole receiver positioned at same depth as source (all dimensions normalised to sphere radius).[22]

DETECTION AND MAPPING OF TUNNELS AND CAVES

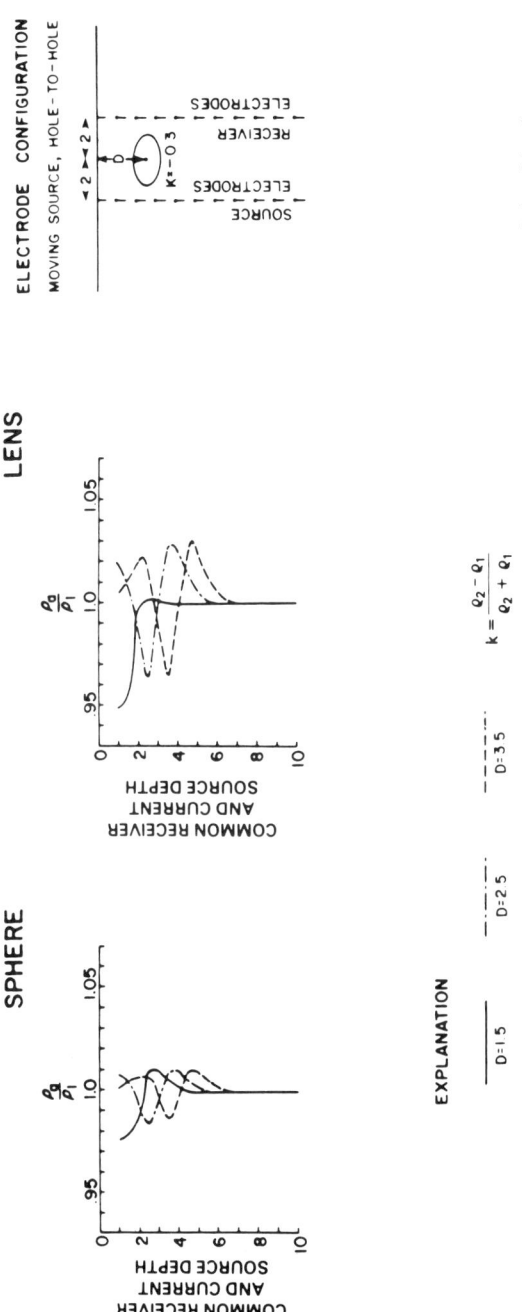

FIG. 35. Normalised apparent resistivity profiles for moving source bipole in one hole and moving receiving bipole at same depth in adjacent hole. Depth of electrodes is as defined in Fig. 33.[22] ($K = (\rho_2 - \rho_1)/(\rho_2 + \rho_1)$.)

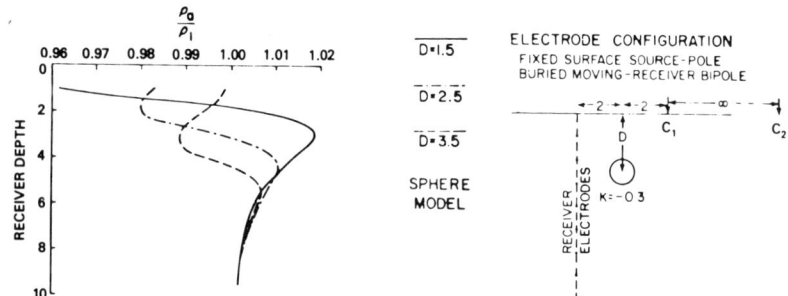

FIG. 36. Normalised apparent resistivity profiles for source current pole (C_1) on surface moving receiving bipole in borehole. Depth of electrodes is as defined in Fig. 33.[22]

anomaly obtained from source and receiver bipoles simultaneously scanned at the same depth in adjacent boreholes. This result indicates that the resistivity anomaly corresponds accurately with the target depth.

When the current source electrodes are located on the surface, the geometric factor does not contain a singularity. Therefore, in this mode of operation, target anomalies are observable without geometric interference. Figure 35 shows the apparent resistivity anomaly for a pole–dipole source in which only the dipole pair is located downhole and the target and

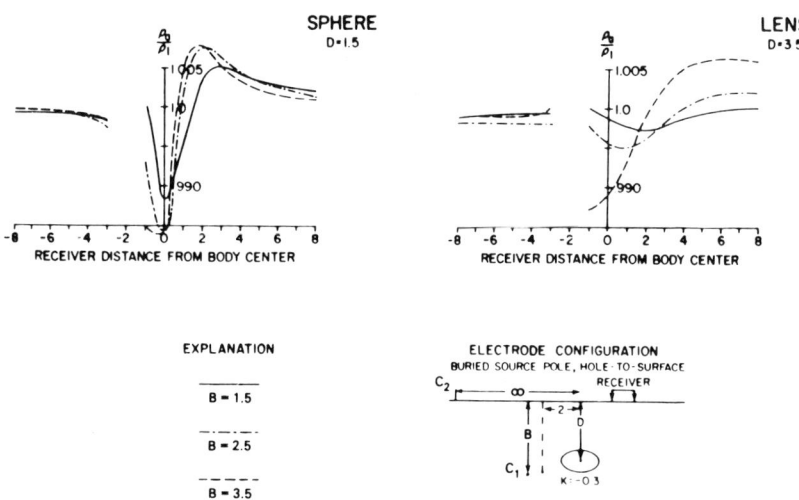

FIG. 37. Normalised apparent resistivity profiles for fixed current source pole in borehole moving receiver bipole on surface. Receiver distance normalised to sphere diameter.[22]

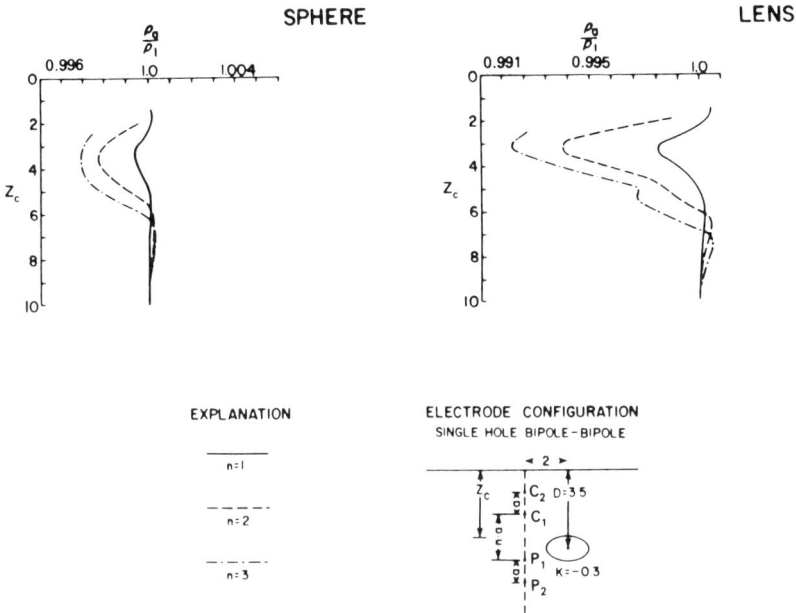

FIG. 38. Normalised apparent resistivity profiles for source current bipole and receiving bipole in single borehole. Dimension a is the sphere diameter and n is an integer factor defining source-to-receiver bipole spacing. Depth Z_c is the depth of the centre of the four-electrode array.[22]

electrode spacing characteristics are the same as those in Fig. 33(b). By comparison, the surface-to-hole target anomaly also occurs at the approximate depth of the target and the peak amplitude of the resistivity anomaly is about twice that observed in the hole-to-hole case.

Reversing the source and receiver electrodes in Fig. 36 to obtain a hole-to-surface measurement arrangement results in target anomalies observable directly over the lateral target position. This is shown in Fig. 37 for three fixed depths of the current source pole. The maximum amplitude sensitivity to the same target shown in Fig. 36 is about 1·5% when the source pole, target body and receiver bipole are in approximate geometric alignment.

The final case investigated by Daniels is a single borehole containing a bipole–bipole electrode array. This arrangement operates in a manner closely corresponding to the surface bipole–bipole array in which the spacing between the source and receiver governs the depth of penetration about the borehole. Figure 38 shows the response to the same target used

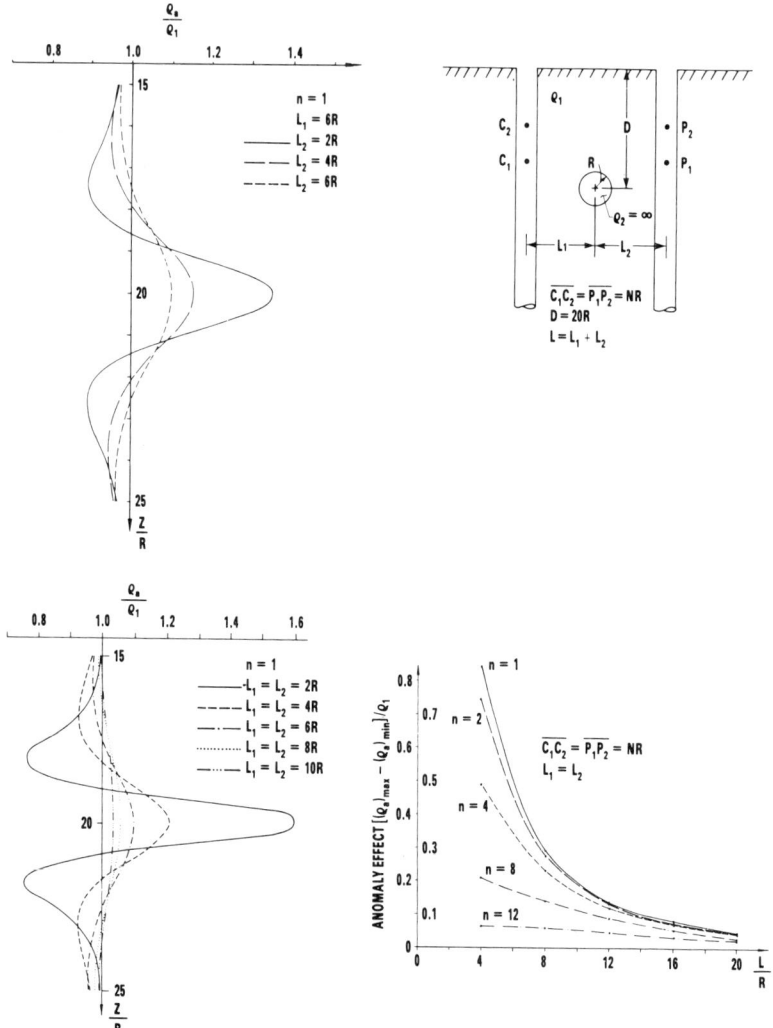

FIG. 39. Normalised two-dimensional apparent resistivity profiles for hole-to-hole bipole–bipole survey of cylindrical tunnel.[43]

earlier, indicating the maximum peak amplitude response of the spherical target to be about 0·5% and about 1% for the lens target.

The Southwest Research Institute has adapted a pole–dipole array to single-borehole operation in a manner similar to the bipole–bipole array described above. This arrangement is being developed for automatic operation using digital data acquisition of spatially overlapping resistivity profiles so that data-processing techniques similar to those described earlier in connection with surface pole–dipole resistivity surveys can be applied. Experimental test results are not yet available from this prototype system.

Militzer et al.[43] present master curves for a variety of borehole resistivity survey arrays following the two-dimensional formulations for cylindrical targets as developed by Parasnis.[44] Although such line electrodes prescribed by the two-dimensional analysis are impractical in actual borehole surveys, the results serve well to compare the relative performance of the various electrode arrangements. In an effort to scale the line electrode results to those obtainable with point source electrodes, Militzer et al.[43] discuss a comparison between the theoretical results for downhole bipole–bipole arrays, representing line electrodes and a tunnel target, and experimental model results obtained using point electrodes and a tunnel target of equivalent geometry. From these results, they conclude that the resistivity anomaly from a long tunnel observed with point electrodes is only about 8% below that for the line electrodes.

Figure 39 shows two-dimensional apparent resistivity anomalies for source and receiver bipole arrays scanned so that the receiver array is always at the same depth as the source array but in boreholes located on opposite sides of a cylindrical tunnel target. These master curves show that anomalous target responses are observable at hole spacings up to about 16–20 tunnel radii with the greatest sensitivity occurring when the source and receiver bipole dimensions $\overline{C_1 C_2}$ and $\overline{P_1 P_2}$ are equal to the tunnel radius. The peak value of the anomaly accurately corresponds with the tunnel depth.

An alternate borehole array described by Militzer et al.[43] is the horizontal bipole array shown in Fig. 40. This arrangement provides a unique indication of the tunnel target depth for separation distances $\overline{C_1 C_2}$ and $\overline{P_1 P_2}$ up to about three times the tunnel radius with maximum sensitivity occurring when $\overline{C_1 C_2} = \overline{P_1 P_2} = 2R$. In practice, the complementary use of the vertical and horizontal bipole configurations for tunnel detection can be readily applied by simply switching the functions of the

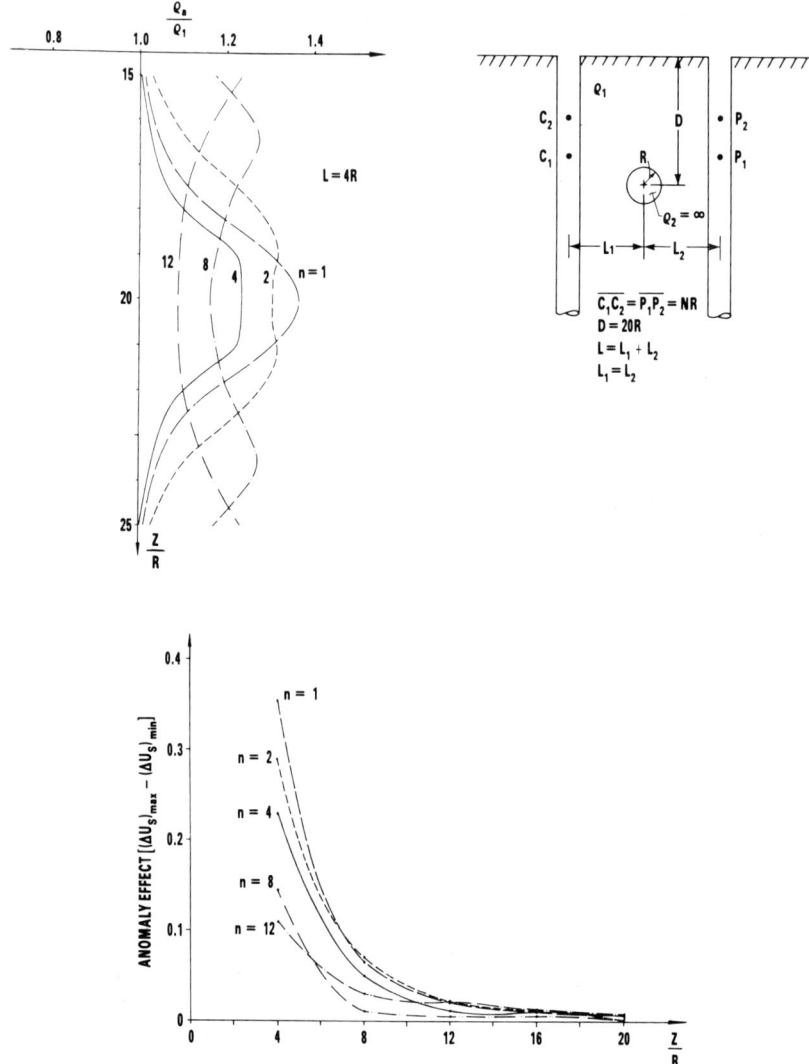

FIG. 40. Normalised two-dimensional resistivity profile for horizontal bipole–bipole array (current applied at C_1C_2; potential measured at P_1P_2).[43]

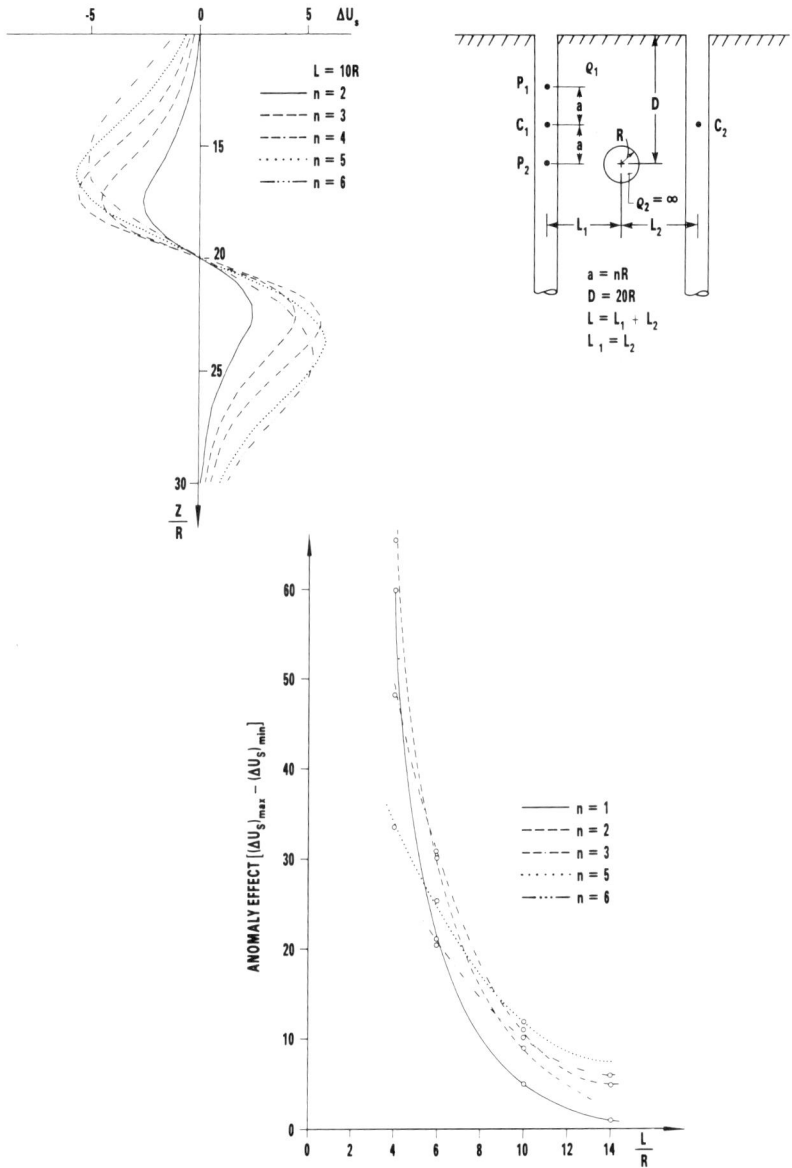

FIG. 41. Normalised two-dimensional apparent resistivity profiles for balanced receiver bipole.[43]

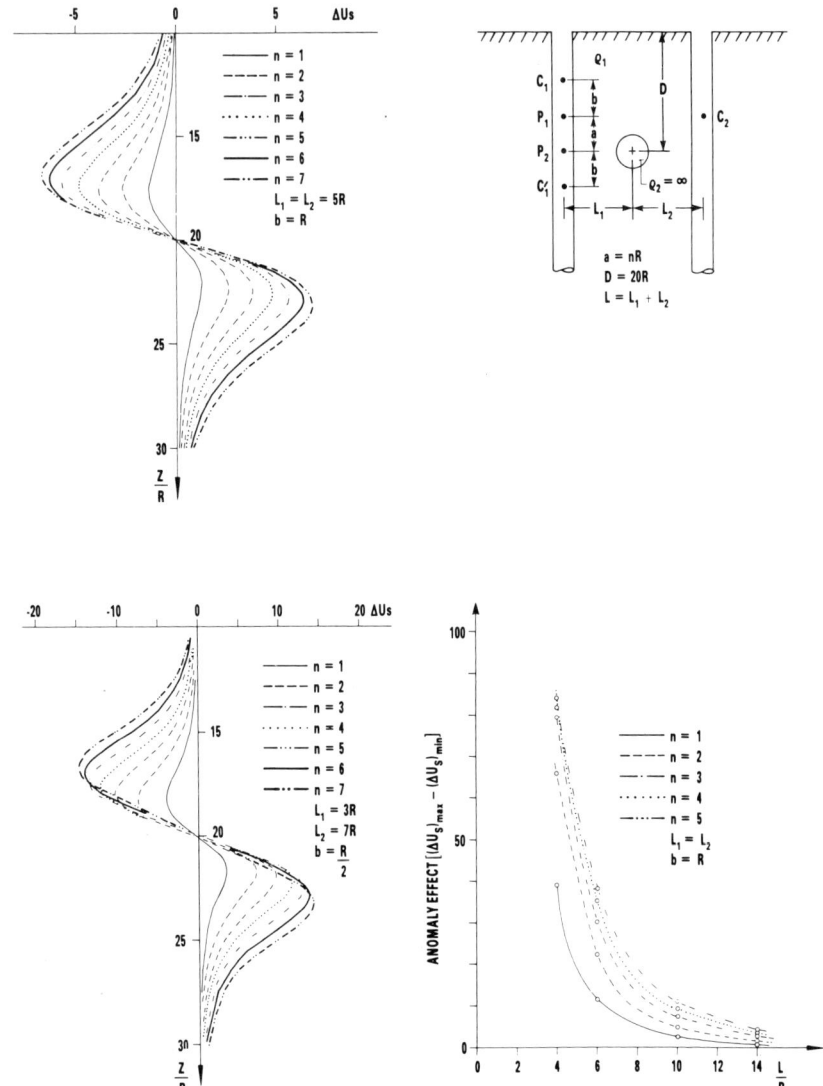

FIG. 42. Normalised two-dimensional apparent resistivity profiles for focused electrode array.[43]

electrodes contained in two borehole cable assemblies (each having a pair of electrodes spaced apart by the anticipated tunnel diameter). The concept of balancing out the primary field is achieved in the potential null arrangement shown in Fig. 41 and in the focusing array shown in Fig. 42. The anomaly response in these curves is the secondary potential produced by the target normalised by the potential that would be obtained in the absence of the target. That is,

$$\Delta U_s = \frac{\pi}{\rho_1 I} U_s \qquad (15)$$

The two types of balanced field arrays exhibit different sensitivity characteristics depending upon the distance between the boreholes. The potential null array dimension $(\overline{P_1 P_2}/R)$ has an optimum value which depends upon the hole spacing L/R, wherein close spacings of $\overline{P_1 P_2}/R \leq 3$ are more favourable for hole spacings up to about $L/R = 5$ and a spacing of about $\overline{P_1 P_2}/R = 4$ is the most sensitive arrangement for $L/R > 8$. In comparison, the focused array is uniformly more sensitive with wider potential electrode spacings, $\overline{P_1 P_2}$, for all hole spacings. Further, the focused array is more sensitive than the potential balancing array for hole spacings up to about $L/R = 10$, beyond which the potential balancing array exhibits a more sensitive threshold of detection. Militzer et al.[43] also show that the anomaly sensitivity observed with the focused array is largely independent of the interelectrode spacing $\overline{C_1 P_1} = \overline{C'_1 P_2}$ except when the boreholes are very close together ($L/R < 4$).

Figure 43 shows computed two-dimensional hole-to-surface resistivity anomalies over a tunnel target for different source bipole depths. The most sensitive response is obtained when the centre of the source bipole is located at the same depth as the tunnel. The peak value of the anomaly occurs at approximately the lateral location of the tunnel relative to the borehole.

Hole-to-hole and hole-to-surface measurement methods can be combined to locate the cavity target with respect to the drill holes. The most favourable electrode configurations for this purpose are the simultaneously scanned vertical bipole–bipole array for the hole-to-hole measurements and the fixed-depth pole–bipole array for the hole-to-surface measurements. The intersection of the horizontal and vertical lines defined by the separately observed anomaly peaks will determine the most probable position of the target.

6.2. Borehole Electromagnetic Techniques

Hole-to-hole electromagnetic wave transmission techniques have been

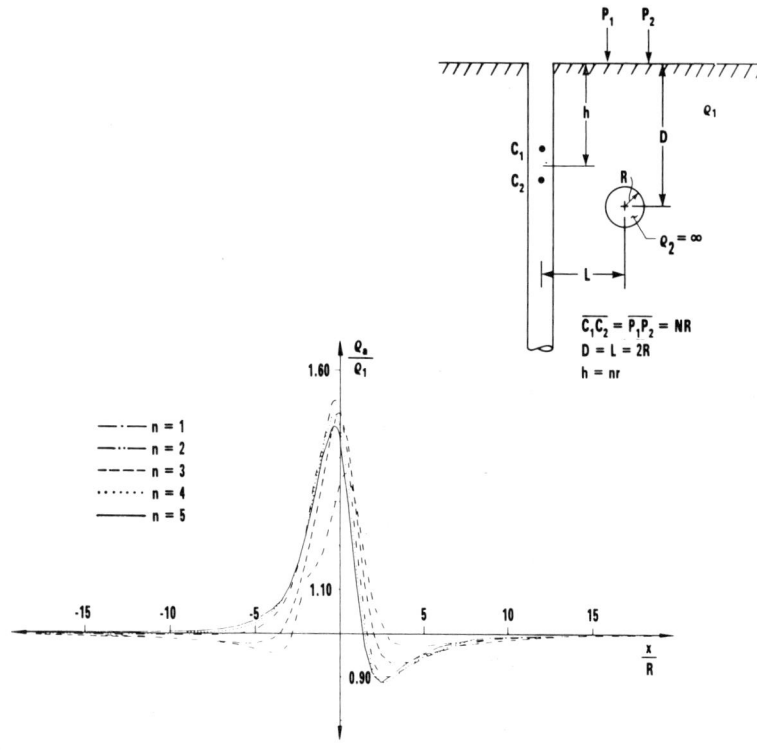

FIG. 43. Normalised two-dimensional apparent resistivity profiles for source current bipole in borehole (depth = D) and moving receiving bipole on surface.[43]

investigated and applied as a means of examining the geological structure between pairs of drill holes. Continuous wave transmission studies conducted at the Lawrence Livermore National Laboratory[45–47] show that forward-scatter signatures from cylindrical voids of arbitrary cross-section can be used effectively in detecting and locating tunnels. Pulse transmission techniques have been applied by the Southwest Research Institute[48,49] to accurately delineate the approximate size and position of tunnels and caves between two drill holes.

Through-transmission signals between two drill holes are governed by the volume distribution of permittivity and electrical conductivity in the intervening geological medium. An air-filled cavity between the holes presents a distinctive contrast with the host medium, giving rise to propagation anomalies that produce characteristic received signal

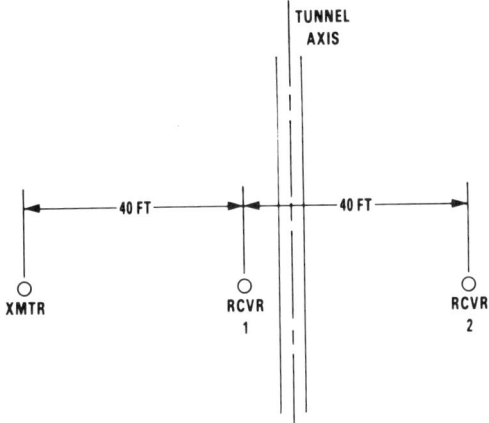

FIG. 44. Borehole positions with respect to tunnel for hole-to-hole electromagnetic tests at Gold Hill, Colorado.[45,47]

signatures. These signatures may be analysed and interpreted in terms of the cavity size and position relative to the hole locations. In the case of air-filled targets in rock media, the contrast in relative dielectric constant is typically in the range 1:9 to 1:16 and the conductivity is many orders of magnitude less than that of the rock. For a mud- or water-filled cavity in rock media, the target contrast in relative dielectric constant is in the range of about 25:16 to 80:9 and the conductivity of the mud or water content is typically 10–100 times greater than the conductivity of the rock (typical rock conductivity is 1–10 S/m in the VHF frequency range).

A controlled field experiment using CW signals at 57 MHz was performed at an abandoned mine adit near Gold Hill, Colorado, by investigators at the Lawrence Livermore National Laboratory.[45,47] The mine adit had a height of 8 ft and a width of 5 ft with a depth to floor level of 84 ft in homogeneous granite rock. The relative dielectric constant and the conductivity of the granite medium were $\varepsilon_r = 12$ and $\delta = 6 \times 10^{-4}$ S/m, respectively. Boreholes were drilled on each side of the adit as shown in Fig. 44 to permit the direct illuminating wave (plus reflections) and the forward-scattered wave to be measured without concern for source antenna coupling along the transmitter borehole.

Figure 45(b) shows the amplitude of the received signals at 57 MHz and the equalised forward-scattered signal observed when the transmitting antenna was stationed at 2-ft depth intervals and the receiving antenna located at the same depth. The signal minima and the 'W' shape of the

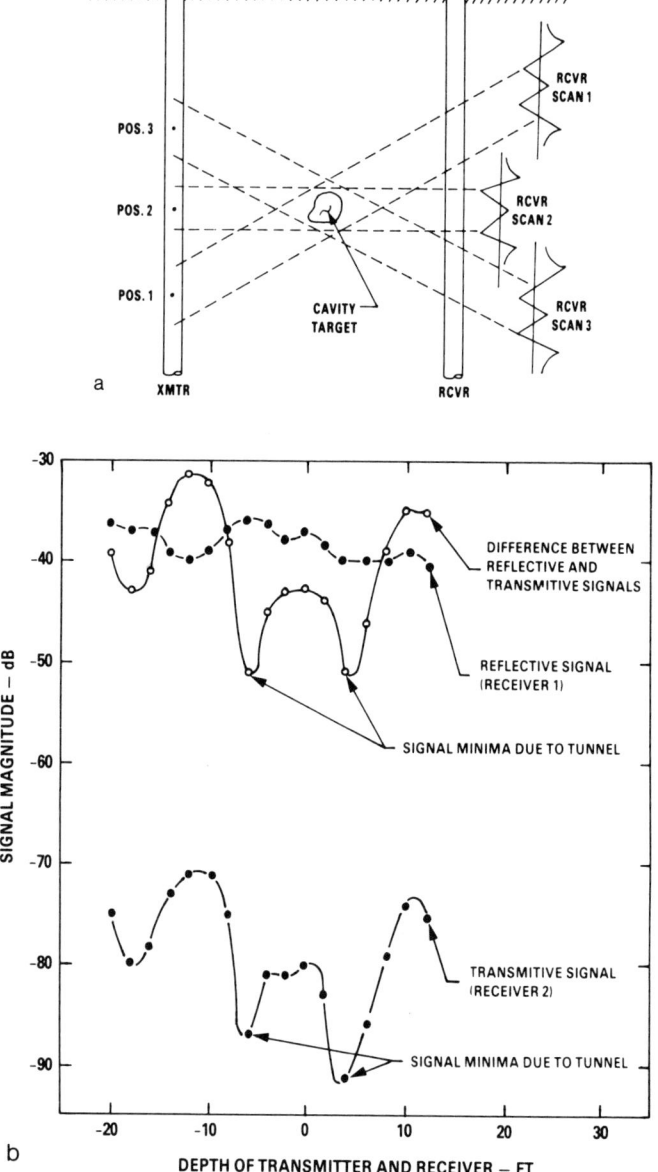

FIG. 45. Hole-to-hole electromagnetic test results at Gold Hill, Colorado. (a) Geometrical method of locating a cavity target between boreholes; (b) experimental amplitude diffraction patterns at 57 MHz.[45,47]

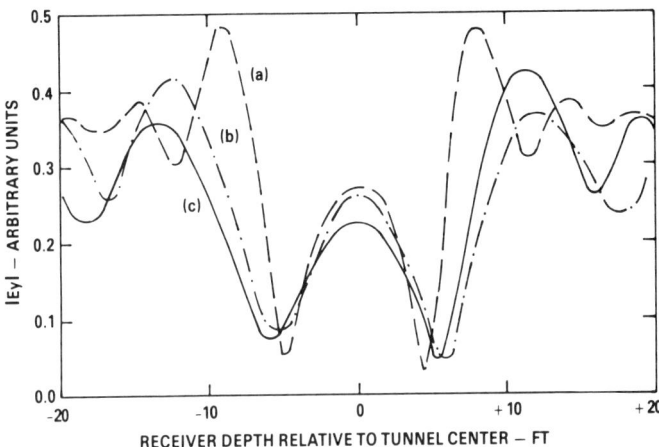

FIG. 46. Computed receiver signals for tunnel target located at (a) 5 ft; (b) 20 ft; (c) 35 ft from receiver borehole. Transmitter-to-receiver distance is 80 ft.[45,47]

transmitted wave diffraction signature are characteristic response features of the tunnel target when illuminated with the 57 MHz signal (wavelength approximately 4·7 ft in granite).

Interpretation of the diffraction signature obtained when the illuminating wavelength is comparable with the tunnel dimension indicates that the two signal minima may be projected onto the transmitter borehole to form spatial limits which contain the tunnel cavity. By offsetting the source and receiver probes in their respective boreholes, the various diffraction signatures may be projected as shown in Fig. 45(a) to determine the tunnel cavity position with respect to the boreholes.

Theoretical computations[46] of the forward-scatter signatures for a two-dimensional model consisting of an infinite line source and an infinite tunnel cavity using a numerical technique developed by Richmond[50] confirm the experimental results shown in Fig. 45.

Calculated 'W'-shaped signatures are shown in Fig. 46 for an 8 × 5-ft tunnel in granite having a slightly arched roof illuminated at different lateral distances by a 57 MHz signal. Two-dimensional numerical computations for other sizes and shapes of tunnels showed that the diffraction signatures from rectangular, circular and triangular cross-sections differ considerably in peak-to-peak amplitudes and in the spacing between their signal minima. However, shape variations such as that between an 8 × 5-ft rectangle and a similar rectangle with an arched roof

FIG. 47. Hole-to-hole ground-penetrating electromagnetic system for tunnel and cavity detection and verification.[48,49]

differ in received signal amplitude only by about 4–5% with the smoother shape exhibiting a slightly reduced signature having essentially the same shape as that of the rectangle.

The hole-to-hole electromagnetic system developed by the Southwest Research Institute[48,49] for cavity detection is shown in Fig. 47. This system provides through-transmission probing of geological materials via a pulse transmitter probe operating in one drill hole and a matching receiver probe operating in an adjacent drill hole typically located 30–100 ft from the source hole. The transmitter radiates a pulse wavelet characterised approximately as one oscillation of a 100 MHz sinewave and having a frequency spectrum covering the 30–300 MHz VHF frequency range. The peak amplitude of the pulse is approximately 550 V corresponding to a peak pulse power of 6000 W at the feed point of the resistance-loaded dipole antenna. The repetition rate of the transmitter is 60 000 pulses/s.

If a subsurface cavity, as interpreted from the results of a resistivity survey, is missed by the first two drill holes, then hole-to-hole electromagnetic tests can be applied. For this purpose, the transmitter and receiver probes are first lowered to the greatest depth of interest and hoisted while operating as a through-transmission system. The radiated source pulse travels through the geological medium between the holes and is detected by the downhole receiver and recorded in full waveform by the surface system.

The hole-to-hole electromagnetic system shown in Fig. 47 consists of a $2\frac{1}{4}$-in diameter transmitter probe 8 ft in length, a similar size receiver probe, a dual-drum wireline winch, a surface control unit and a digital magnetic tape recording system. The fibreglass borehole probes consist of tubular resistively-loaded dipole antennae containing the respective pulse transmitter and receiver electronic modules. Time synchronisation of the receiver with respect to the transmitter pulse is provided to allow time-domain sampling of the received signals. This sampling system translates RF signals received in the 30–300 MHz VHF downhole receiver frequency range to replica waveforms in the 600–6000 Hz audiofrequency range for transmission uphole via conventional armoured logging cable. The surface control unit powers and operates the downhole probes and accepts the time-domain-sampled pulse waveforms for conversion to digital format for recording on magnetic tape. The dual-drum wireline winch allows the two probes to be independently positioned in their respective drill holes and hoisted either independently or simultaneously to any desired relative positions.

The propagation velocity and attenuation of the transmitted pulse is governed by the dielectric and conductivity properties of the medium. For example, electromagnetic waves travelling along ray paths passing through or diffracted by an air-filled cavity travel slightly faster than those travelling through the surrounding host ground. Thus, by hoisting the probes so that the transmitter is at the same depth as the receiver, anomalous pulse propagation conditions caused by a cavity can be readily observed at the target depth. By offsetting the transmitter and receiver probes at known distances along their respective boreholes and rescanning past the target, the approximate location of the cavity can be determined by simple graphic projections of the common cavity anomaly. Figure 48 illustrates the probe arrangement used in locating the cavity position relative to the two boreholes.

Performance of the hole-to-hole electromagnetic system may be defined by the spatial resolution of the pulse waveform in the drilled geological medium and its typical limitations in hole-separation distance. The practical maximum hole spacing experienced with the VHF pulse system is about 30 m in rock media and the corresponding spatial dimension of the pulse waveform is about 1 m. In more dissipative geological materials such as highly porous water-saturated rock or soils, the maximum usable hole spacing may be limited to only a few metres.

Figures 49–51 show the results of field tests performed at Medford Cave, Florida, using a hole-to-hole electromagnetic probing system. The test

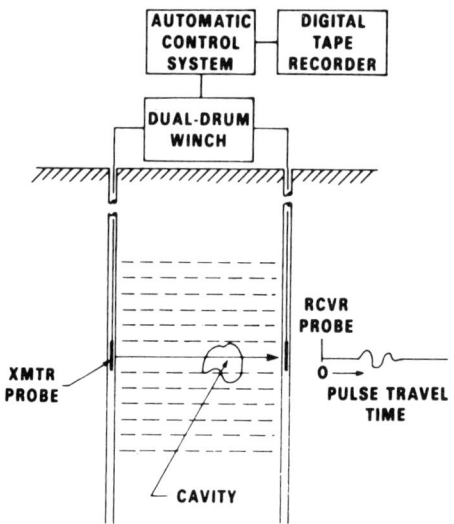

FIG. 48. Hole-to-hole scanning arrangement for electromagnetic tests at Medford Cave, Florida.[48]

layout at Medford Cave is shown in Fig. 49(a). Eight drill holes were located at various positions relative to the openings of the shallow limestone solution cave. The test results were obtained using only horizontal propagation paths between selected pairs of holes. Tests between holes C4 and C5 show the general uniformity of the electromagnetic propagation velocity in the limestone rock where no cavities exist.

Tests between holes C5 and C3 indicate a small thin cavity at a depth of about 7 m below the surface. Tests between holes C2 and C3 show a larger room of the cave at a depth of about 9 m. Tests between holes C2 and C5 show the combined effects of the two previous cavities as observed between more widely spaced holes.

The farthest spaced holes available at the test site were holes C4 and C8 which were about 24 m apart. The tests using these holes show the main room of the cave at a depth of about 7–8 m. The signal-to-noise ratio indicated in these data is sufficiently high that this cave target could very likely be detected or verified at a hole separation of ≥ 50 m.

6.3. Borehole Seismic Techniques

Seismic techniques have been investigated and developed utilising hole-to-hole and hole-to-surface transmission of compressional waves to detect

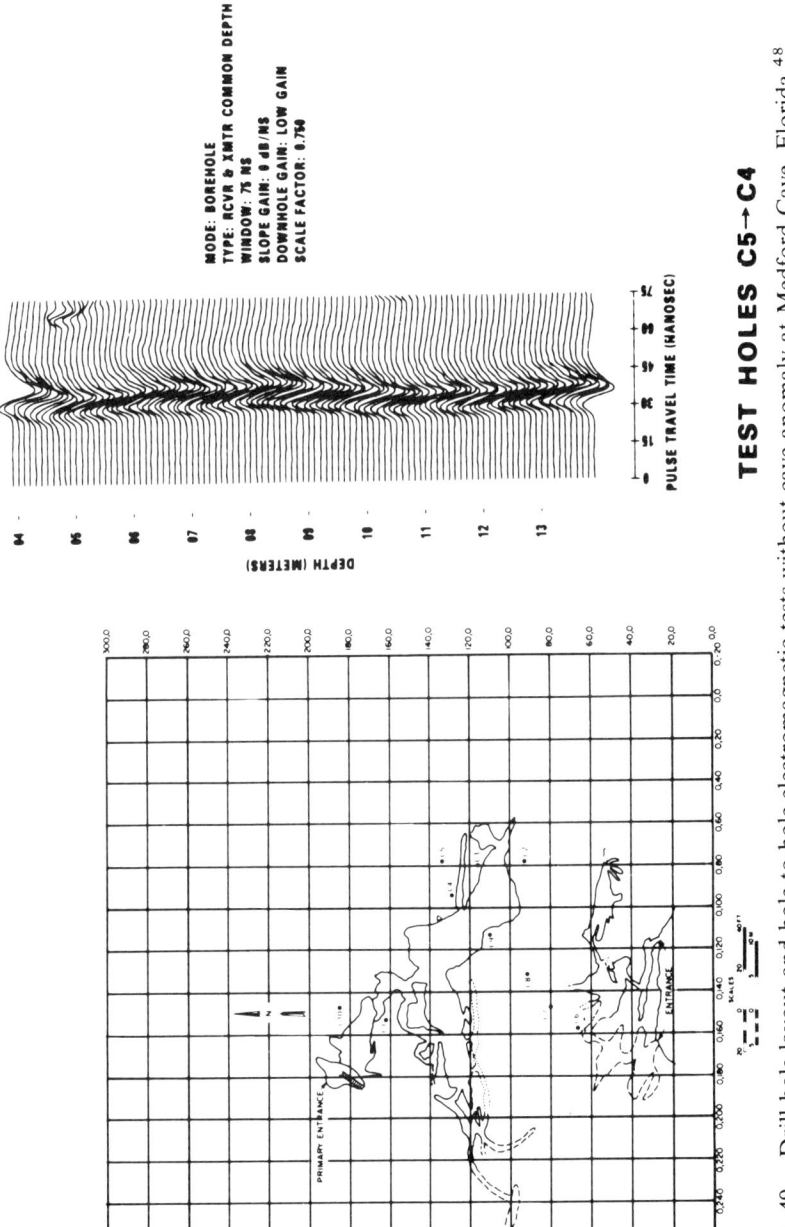

Fig. 49. Drill hole layout and hole-to-hole electromagnetic tests without cave anomaly at Medford Cave, Florida.[48]

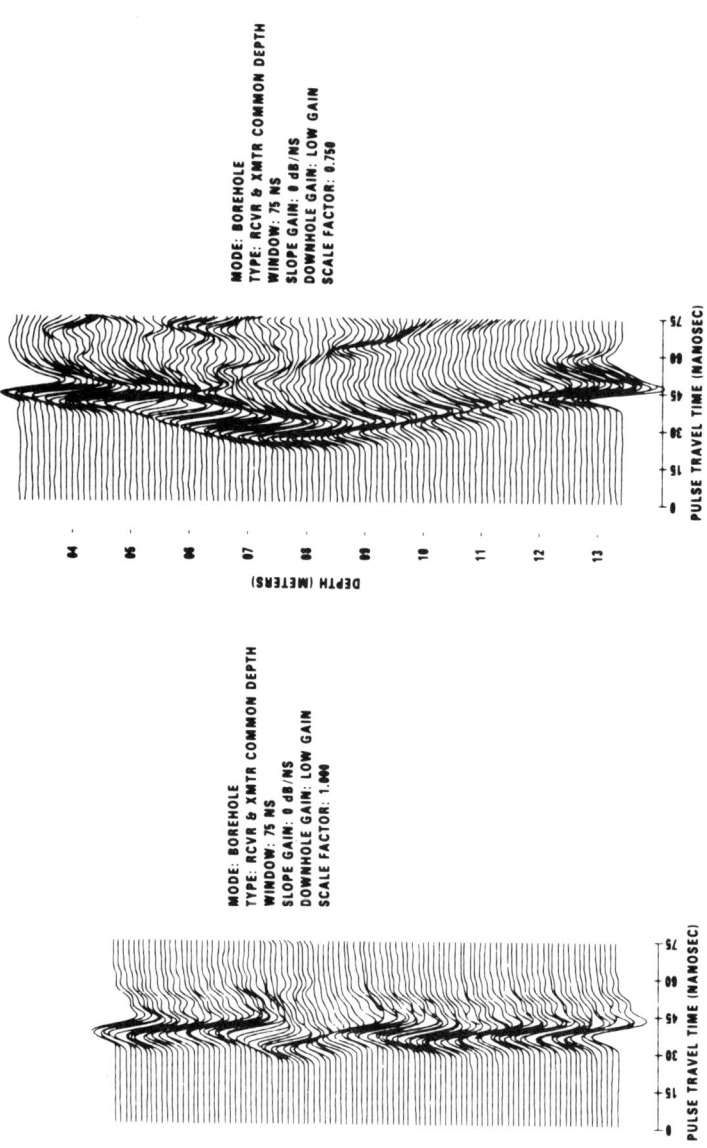

FIG. 50. Hole-to-hole electromagnetic test results from Medford Cave, Florida.[48]

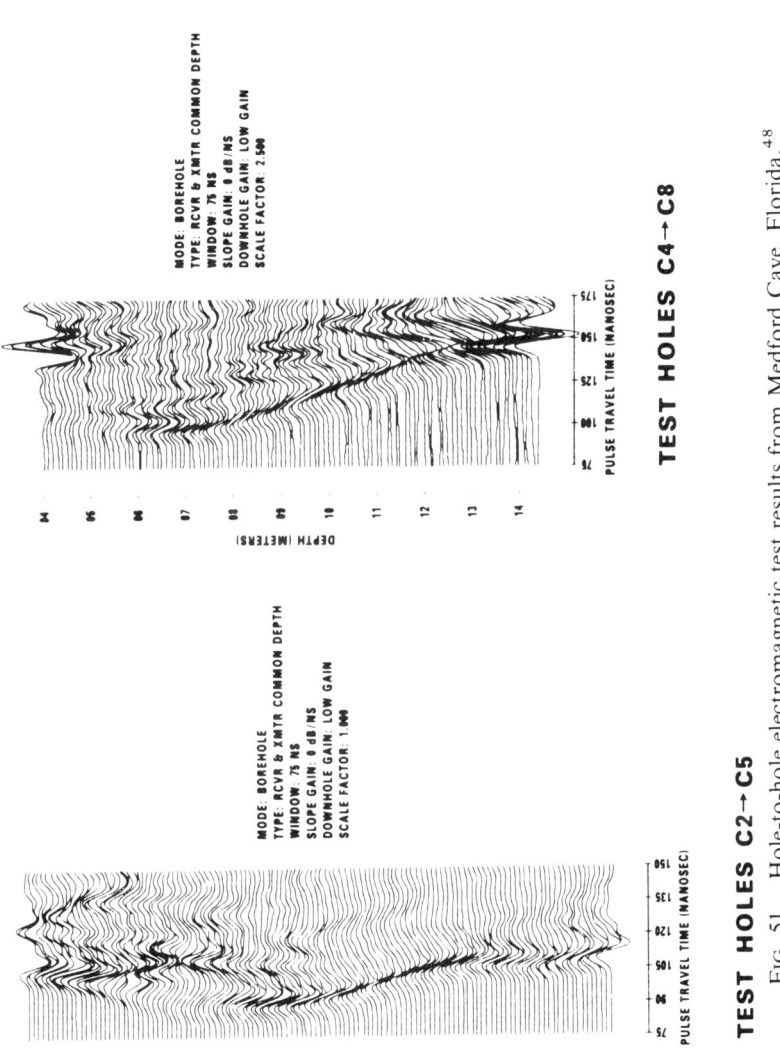

FIG. 51. Hole-to-hole electromagnetic test results from Medford Cave, Florida.[48]

cavities by their anomalous forward-scattering and wave-diffraction effects. For relatively deep-probing applications, hole-to-hole seismic signals received in one vertical drill hole as a result of a source operating in an adjacent hole can be analysed to interpret the inhomogeneous conditions in the medium located between the holes. In this regard, seismic wave techniques are applicable in the same ways as described for electromagnetic waves. The diffraction patterns of interest are of the characteristic 'W' shape illustrated earlier and are oriented in the vertical plane containing the source and receiver boreholes. Therefore, by utilising several pairs of holes in differently oriented planes, the approximate three-dimensional size and shape of a localised cavity or other form of seismic inhomogeneity can be determined.

When a localised cavity target is near the surface but too deep to be detected by the Rayleigh wave seismic technique described earlier, the source alone may be operated in a drill hole to produce observable amplitude diffraction patterns in either compressional or shear waves observed at the surface. Angular scans using surface-coupled detectors at a uniform radial distance from the downhole source transducer will yield 'W'-shaped amplitude patterns for diffracting targets whose diameters are in the range of about one-quarter to three wavelengths when they are located in the propagation path between the source and receiver scan layout. By varying the source depth and with the use of two or more source drill holes, the location of the cavity target can be determined relative to the drill holes and the surface detector layout positions. Similar results may be obtained using a single detector in the drill hole and a larger surface source at the scan positions.

Borehole seismic methods of this type have been demonstrated in practice by Dresen[51] and Lange[52] in reference to the hole-to-hole technique and Dresen and coworkers[53-55] using the hole-to-surface technique. These efforts have included physical scale model studies[51,52] which have produced distinctive and quantitative interpretation techniques for deriving the size and location of cavity targets from the features of the 'W'-shaped amplitude diffraction pattern. Practical applications of the techniques have been demonstrated to detect shallow abandoned mine entry shafts and subsurface cavities[55] in Europe.

Figure 52 shows (in plan) the general hole-to-hole seismic method as applied to the detection of a tunnel target at depth and the arrangement also represents an approximate model of the hole-to-surface measurement method. The fracture zone which surrounds many subsurface tunnels is an important part of its geometry in that the elastic moduli of the nearby medium are modified from those of the undisturbed medium because of

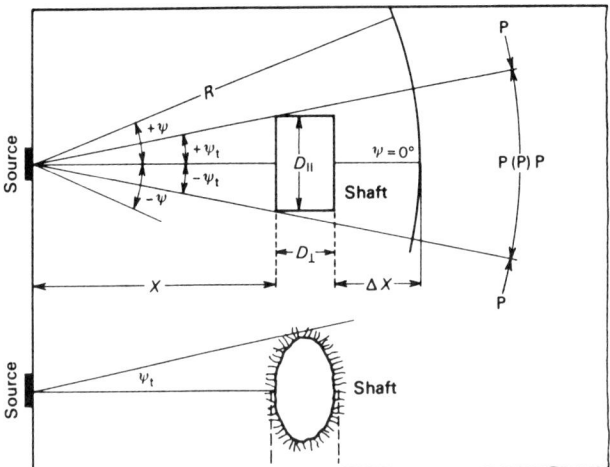

FIG. 52. Two-dimensional model simulation arrangement used in the investigation of hole-to-hole seismic techniques.[54]

local microfracturing and changes in the amount of pore fluid contained near the cavity boundary. Forward scattering from the cavity produces amplitude and travel-time anomalies when observed in the plane of the drill holes.

In addition, the frequency-selective scattering characteristics of the cavity obstacle give rise to a principal frequency component in the received signal at various scan angles which may further aid in locating the cavity and possibly provide supplementary information on the general shape and roughness of the cavity geometry. This frequency-selective effect observed in the forward-scattered signal may be explained in simple terms as the constructive interference between diffracted waves travelling around the cavity in different directions and is a combined result of variations in effective path length and travel time versus wavelength in the vicinity of the cavity.

Figure 53 shows typical model test results obtained from an arrangement such as that in Fig. 52 in which several different shaft cross-sectional shapes were investigated. The amplitude diffraction curves show very similar characteristics indicating that various sizes of shafts which cast the same shadow zone at the receiver scan path cannot be uniquely identified from one another.

The influence of the fracture zone that surrounds practical mine shafts tends to aid in their detection in that the effective size of the cavity is larger. Figure 54 shows a comparison between the frequency dependency of the

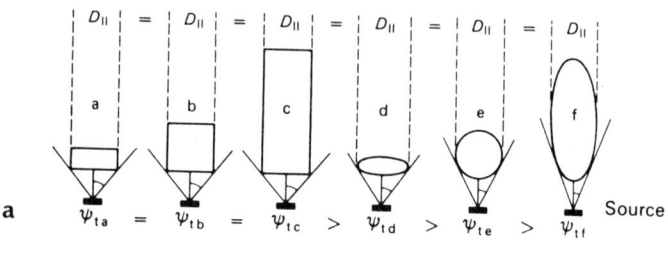

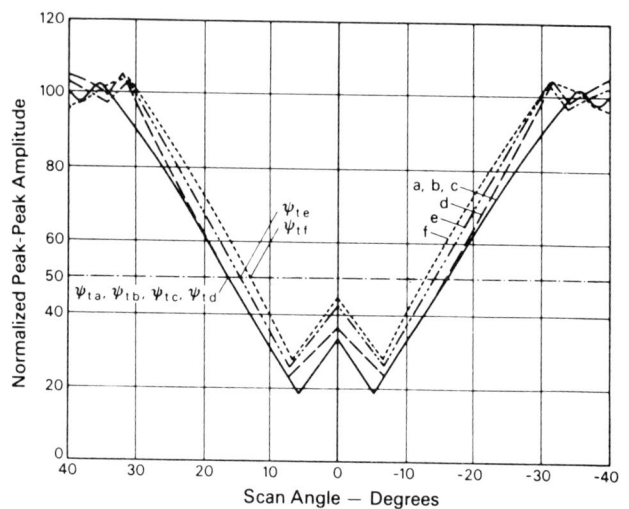

FIG. 53. Two-dimensional model test results obtained with six idealised mine shaft cavity shapes. (a) Typical shaft cross-sections tested in model form; (b) amplitude diffraction patterns for several typical shaft cross-sections.[54]

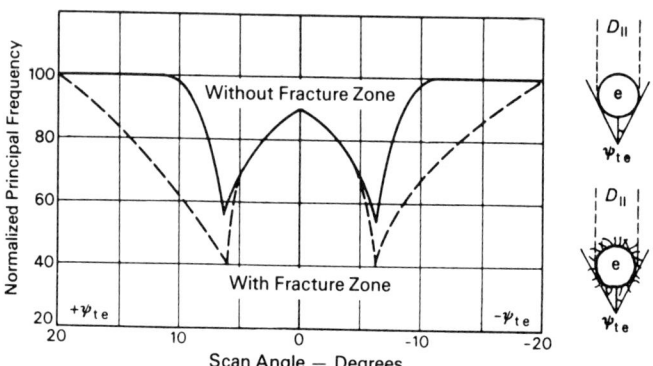

FIG. 54. Effect of a surrounding fracture zone on the frequency-dependent scattering characteristic of a circular cavity.[54]

forward-scattered compressional wave for smooth and fracture-bounded cavity models. Although the two cases have nearly equal shadow zone angles in the test results shown, the principal frequency component observed in the signal scattered by the fracture-zone cavity has a characteristically different pattern which, in general, will make the shaft more readily detectable.

The hole-to-surface method has been successfully demonstrated in practice using a weight-drop source on the surface and an accelerometer detector in the borehole. Figure 55 shows the scan layout around Shaft I

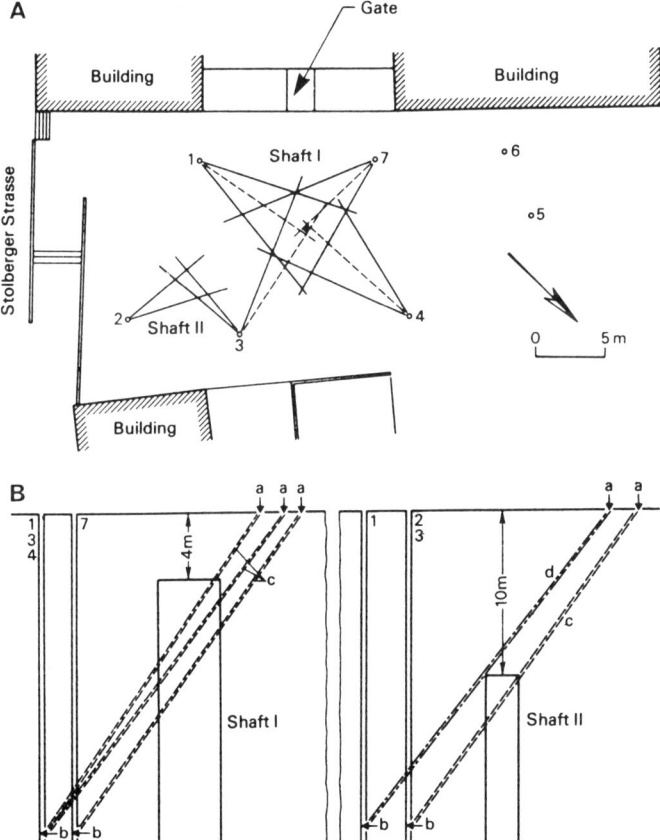

FIG. 55. Hole-to-surface test layout at the Herren-Kunst Mine Shaft. A. Surface scan layouts and borehole locations. B. Cross-section showing subsurface shafts and (a) weight-drop source and (b) borehole accelerometer locations.[54]

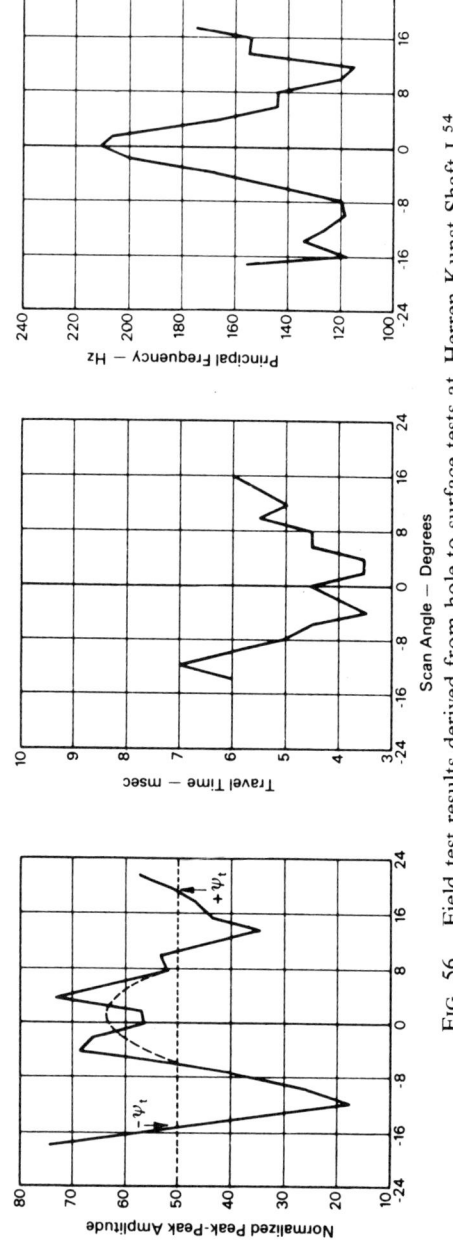

FIG. 56. Field test results derived from hole-to-surface tests at Herren-Kunst Shaft I.[54]

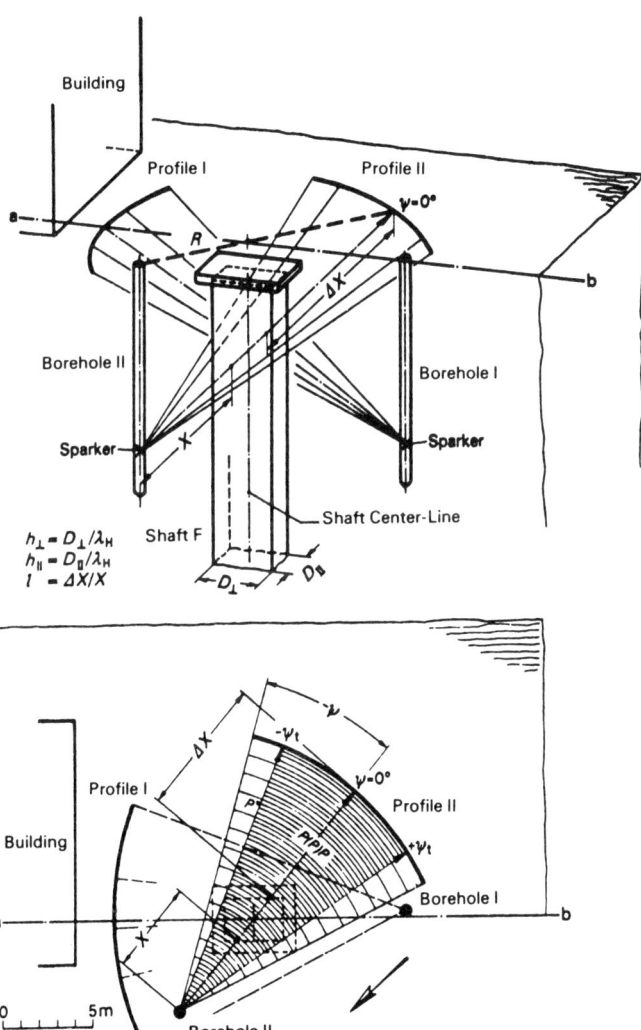

FIG. 57. Diagrammatic sketch of the hole-to-surface test layout at Shaft F, Wetter-on-the-Ruhr, FRG.[53]

and Shaft II at the Herren-Kunst Mine in Eschweiler, FRG. Test data collected in borehole 4 at Shaft I are illustrated in Fig. 56. These results show all of the distinctive forward-scattering effects observed in the model studies. The tangential ray angles $\pm \psi_t$ determined from the 50% amplitude value of the amplitude diffraction pattern establishes a measure of the projected shaft width. Thus, by observing this shadow boundary angle

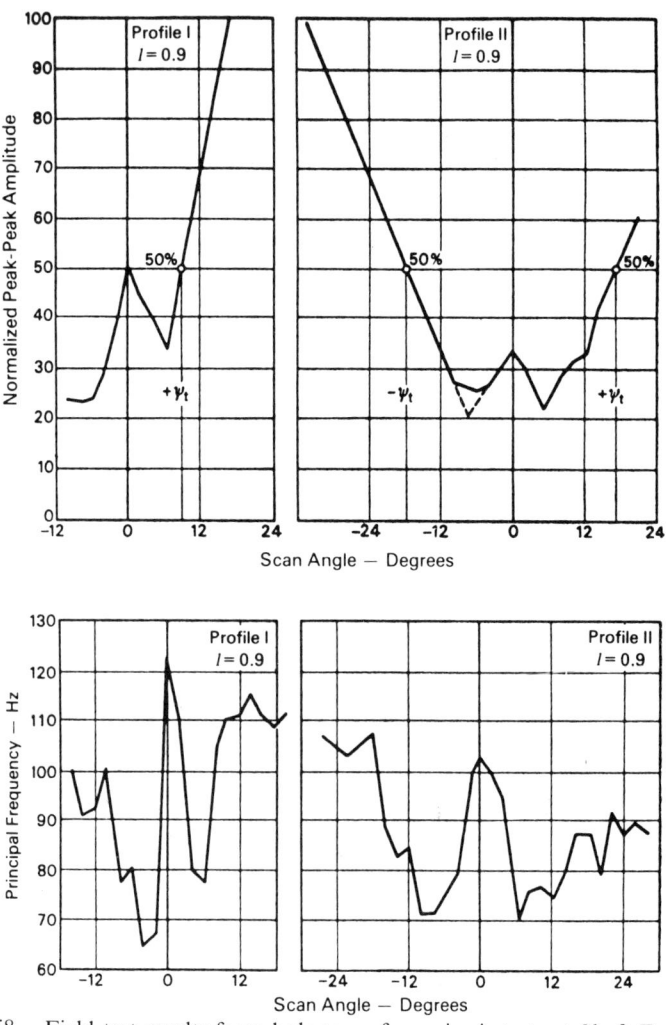

FIG. 58. Field test results from hole-to-surface seismic tests at Shaft F, Wetter-on-the-Ruhr, FRG.[53]

from the several boreholes (Nos. 1, 4 and 7 in Fig. 55A), the location and approximate size of the shaft may be determined.

Another example of the hole-to-surface seismic technique is shown in Fig. 57 for measurements at Shaft F in Wetter-on-the-Ruhr, FRG. In this case a spark discharge source was used in the boreholes and a geophone on the surface. The amplitude, travel time and principal frequency anomalies derived from the measured data are shown in Fig. 58. The half-amplitude angular widths of the amplitude diffraction patterns are plotted in Fig. 59 indicating the excellent predicted shaft location in comparison with the actual location.

6.4. Borehole Gravity Methods

The Lacoste and Romberg borehole gravity meter is capable of providing gravity measurements in drill holes with an instrument error of less than $\pm 5\,\mu$gals. Therefore, if a tunnel or cave target is too deep to be detected using surface gravity methods, it may still be detectable at depth from a borehole located within reasonable proximity. In general, the borehole gravity measurements are affected less by topographical noise and surface vibration effects than by other error parameters such as uncertainties in exact instrument depth and borehole deviations from an assumed vertical path. LaFehr et al.[56] have discussed the application of borehole gravity methods to the detection of tunnels. They state that the nominal change in gravity versus depth is in the range of 65–165 μgal/m, depending upon the

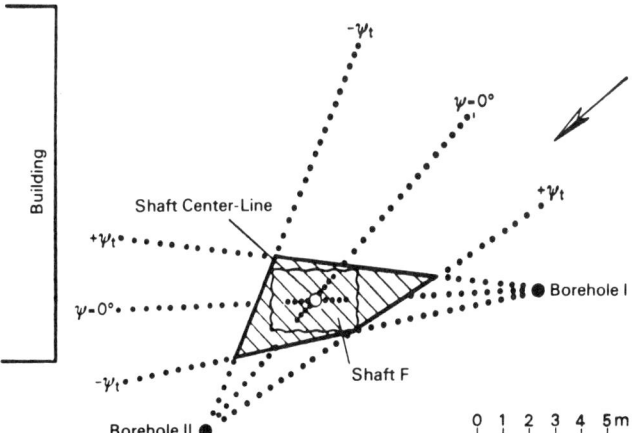

FIG. 59. Comparison of experimental seismic detection and location of Shaft F with actual location.[53]

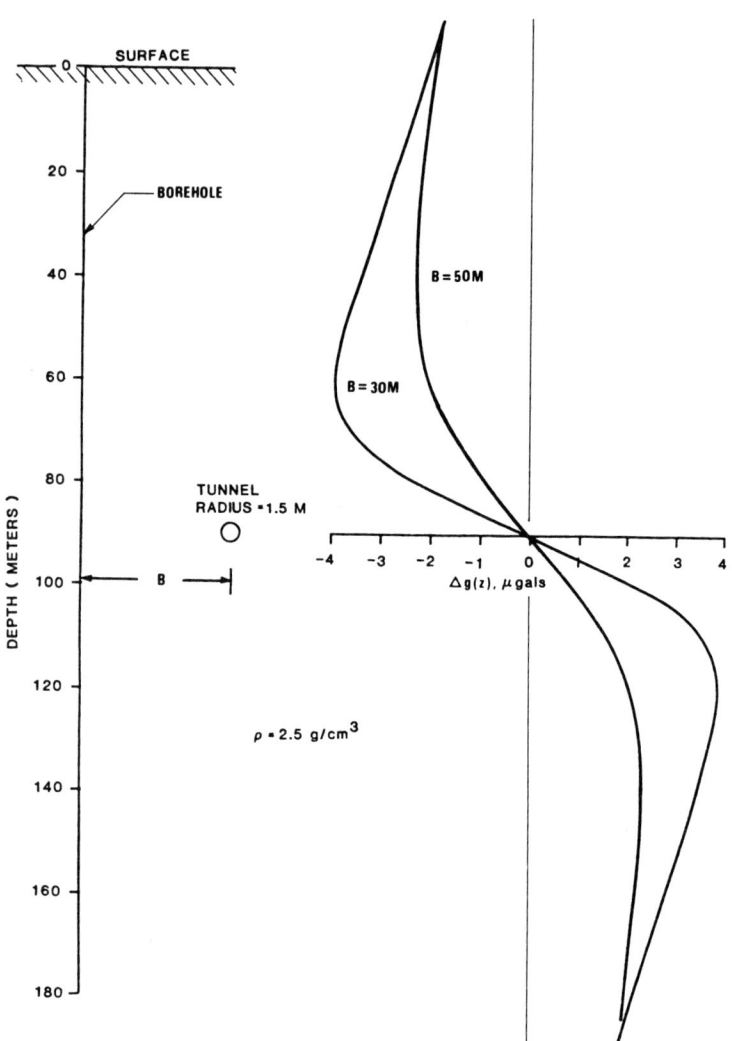

FIG. 60. Calculated borehole gravity effect produced by an air-filled cylindrical tunnel.[56]

geological formation being probed. Therefore, particular care must be taken to measure the instrument depth within a relative accuracy of about $\lesssim 2\%$ at adjacent measurement stations in order to hold the positional error below the intrinsic measurement error range of the instrument.

The anomalous gravity effect characterising an infinite cylindrical tunnel target in a homogeneous medium when measured by a borehole gravity meter scanned along a vertical drill hole is shown in Fig. 60. For a tunnel target having a radius of 1·5 m and an offset distance of 30 m away from the borehole, the peak-to-peak gravity effect is 8 μgals. Since this maximum anomalous gravity effect is inversely proportional to the cavity offset distance, a practical detection distance between the tunnel and the borehole is in the range of 10–15 m, depending upon the homogeneity of the geological host medium. To date, no definitive field tests have been conducted to demonstrate the practicability or success of borehole gravity detection of cavities in the field. At present, this method is limited in application because of the shortage of borehole gravimeters readily available for use in tunnel and cave surveys.

7. CONCLUSIONS

Although the subject of tunnel and cave detection and mapping is not particularly new, concerted efforts have been applied to a variety of such problems during recent years. The applications of interest have typically involved relatively small or deep targets which demanded that the geophysical detection technique be extended to its maximum limits of performance in sensitivity and spatial resolution. One result of these efforts has been the investigative study of the threshold detection capabilities of various surface and borehole geophysical techniques and the adaptation of the more promising methods to specific practical needs.

Advances in both field instrumentation and in data-processing capabilities have contributed to the success of several applicable techniques from the viewpoint of providing the practical means for gathering high-resolution survey data as well as in enhancing and extracting the target signals from the noise imposed by heterogeneous geological background conditions.

Emphasis on development and application of geophysical methods in tunnel and cave detection has resulted from the need to search for and locate such anomalous conditions without the impractical expense of extensive drilling programmes capable of providing comparable detection reliability and resolution. Engineering surveys of critical foundations such as those for nuclear reactor installations and water impounding dams are important ongoing applications which can be expected to demand high-performance capabilities from a variety of practical ground-probing techniques. Ground subsidence problems stemming from shallow underground mine workings and from incipient sinkhole conditions in karst terrain also support the need for practical geophysical methodology capable of minimising the increasing costs of conventional drilling and test boring surveys.

The particular technical approaches to these applications discussed earlier are presently the most effective and well-developed methods. Each method has additional potential for some degree of improvement as the demands for greater performance capabilities dictate. Indeed, the engineering and geotechnical requirements characterising the needs for such high-resolution search and verification ground-probing techniques will continue to encourage advancements in the state of the art. Predictably, such demands are not expected to be satisfied until indirect ground-probing methods are available which not only reveal high contrast anomalies such as cavities but reach the limit of being able to provide quantitative information on the engineering parameters of the ground. Such a capability could greatly improve the reliable design and construction of underground projects, facilitate the evaluation of the load bearing strength and stability of surface foundations, and permit specialised probing and remote sensing of detailed geological structures in underground mining.

Some of the advancements that may be anticipated in the near-future include the prospect of imaging the geological structure being probed in the same way that computerised tomography provides cross-sectional X-ray images in medical applications. This concept is primarily applicable to the borehole electromagnetic and seismic wave techniques discussed earlier; however, certain analogies of application may permit the method to be adapted to those geophysical methods which are based upon potential theory. Dines and Lytle[57] have discussed the applicability of tomographical techniques to certain geophysical problems.

A subtlety underlying all of the geophysical techniques presented in connection with tunnel and cave detection is the fact that each method responds to different physical contrasts and parameters of the target. The possibility therefore arises that more than one ground-probing technique can be combined to provide improved target detection reliability and interpreted physical detail. By integrating two or more sensing techniques together via certain theoretical and empirical linkages, such as has been done in well-logging technology, the potentials of deeper penetrating geophysical probing methods will be more completely fulfilled.

REFERENCES

1. PALMER, L. S. and HOUGH, J. M. (1953) Geoelectrical resistance measurements. *The Mining Mag.* (*London*) **88**, 16–22.
2. PALMER, L. S. (1954) Location of subterranean cavities by geoelectrical methods. *The Mining Mag.* (*London*) **91**, 131–47.
3. PALMER, L. S. (1959) Examples of geoelectric surveys. *Proc. Institution of Electrical Engineers* **106A**, 231–44.
4. TRATMAN, E. K. *et al.* (1963) Reports on the investigation of Pen Park Hole, Bristol. In: *Geoelectric Survey and Investigation*, ed. L. S. Palmer, Cave Research Group, Vol. 12, Chap. 2, pp. 1–54.
5. BRISTOW, C. M. (1966) A new graphical resistivity technique for detecting air-filled cavities. *Studies in Speleology* **1**(4), 204–27.
6. BATES, E. R. (1973) Detection of subsurface cavities. Misc. Paper S-73-40, US Army Engineer Waterways Experiment Station, Vicksburg, Miss.
7. FOUNTAIN, L. S., HERZIG, F. X. and OWEN, T. E. (1975) Detection of subsurface cavities by surface remote sensing techniques. Report No. FHWA-RD-75-80, FHWA Contract No. DOT-FH-11-8496.
8. FOUNTAIN, L. S. (1975) Evaluation of high-resolution earth resistivity measurement techniques for detecting subsurface cavities in a granite environment. Final Tech. Rep., US Army Mobility Equipment R&D Command, Fort Belvoir, Virginia, Contract DAAG53-75-C-0213.
9. FOUNTAIN, L. S. (1981) A chronology of developments in tunnel detection. *Proc. Symposium on Tunnel Detection*, Colorado School of Mines, Golden, Colorado (21–23 July), sponsored by US Army Mobility Equipment R&D Command, Fort Belvoir, Virginia.
10. PETERS, W. R. (1981) Use of an automatic earth resistivity system for the detection of tunnels. *Proc. Symposium on Tunnel Detection*, Colorado School of Mines, Golden, Colorado (21–23 July), sponsored by US Army Mobility Equipment R&D Command, Fort Belvoir, Virginia.
11. PETERS, W. R. and BURDICK, R. G. (1981) Use of an automatic earth resistivity system for detection of abandoned mine workings. Preprint No. 81–89, AIME Annual Meeting, Chicago, Illinois.
12. SPIEGEL, R. J., STURDIVANT, V. R. and OWEN, T. E. (1980) Modeling resistivity anomalies from localised voids under irregular terrain. *Geophysics* **45**, 1164–83.
13. HABBERJAM, G. M. (1969) The location of spherical cavities using a tripotential resistivity technique. *Geophysics* **34**, 780–4.
14. VAN NOSTRAND, R. G. (1953) Limitations on resistivity methods as inferred from the buried sphere problem. *Geophysics* **18**(2), 423–33.
15. MILITZER, H., RÖSLER, R. and LÖSCH, W. (1979) Theoretical and experimental

investigations for cavity research with geoelectrical resistivity methods. *Geophys. Prosp.* **27**(3), 640–52. (Also: (1977) Theoretische Modellkures zum Geoelektrischen Hohlraumnachweis. VEB Bus Welzow, Grossrauschen, DDR.
16. LÖSCH, W., MILITZER, H. and RÖSLER, R. (1976) Die Anwendung von geoelektrischen Messverfahren mit fokussierendem Effekt zur Ortung oberflächennaker Hohlräume. *Neue Bergbautechnik* **6**(12), 898–905.
17. VAN NOSTRAND, R. G. and COOK, K. L. (1966) Interpretation of resistivity data. US Geolog. Survey Professional Paper No. 499, p. 310.
18. LIPSKAYA, N. V. (1953) Anomaly field produced by a local heterogeneity of finite electro-conductivity. *Akad. Nauk. SSR Izv. ser. Geophyz.* **6**, 514–22.
19. SINGH, S. K. and ESPINDOLA, J. M. (1976) Apparent resistivity of a perfectly conducting sphere buried in a half-space. *Geophysics* **41**, 742–51.
20. LARGE, D. B. (1971) Electric potential near a spherical body in a conducting half-space. *Geophysics* **36**, 763–7.
21. SNYDER, D. D. and MERKEL, R. M. (1973) Analytic models for the interpretation of electrical surveys using buried current electrodes. *Geophysics* **38**, 513–29.
22. DANIELS, J. J. (1977) Three-dimensional resistivity and induced-polarization modeling using buried electrodes. *Geophysics* **42**, 1006–19.
23. COOK, J. C. (1965) Seismic mapping of underground cavities using reflection amplitudes. *Geophysics* **30**(4), 527–38.
24. OWEN, T. E. and DARILEK, G. T. (1976) High-resolution seismic reflection measurements for tunnel detection. Interim Tech. Rep., US Army Mobility Equipment R & D Command, Fort Belvoir, Virginia, Contract DAAG53-76-C-0160.
25. RECHTIEN, R. H., STEWART, D. M. and CAVANAUGH, T. (1976) Seismic detection of subterranean cavities. Final Tech. Rep., US Army Research Office Contract DAAG29-76-G-0006.
26. RUSKY, F. and SNYDER, L. (1981) Seismic and resistivity techniques for locating abandoned coal mine workings. Paper E2.2 presented at 51st Meeting of the SEG, Los Angeles, California.
27. TURPENING, R. M. and ADAMS, J. W. (1975) Feasibility study of detecting voids by shear wave refraction methods. Final Tech. Rep., Bureau of Mines Contract No. H0242030.
28. DRESEN, L. and HSIEH, C.-H. (1979) Ortung verlassener Schächte mit Hilfe von Rayleigh-Wellen. *Glückauf-Forsch.-H.* **40**, 190–8.
29. COLLEY, G. C. (1963) The detection of caves by gravity measurements. *Geophys. Prosp.* **II**, 1–9.
30. ARZI, A. A. (1975) Microgravimetry for engineering applications. *Geophys. Prosp.* **23**, 408–25.
31. FAJKLEWICZ, Z. J. (1976) Gravity vertical gradient measurements for the detection of small geological and anthropogenic forms. *Geophysics* **41**, 1016–30; and related Discussions in (1977) *Geophysics* **42**, 872–6, 1066–9 and 1484–5.
32. FAJKLEWICZ, Z. J. (1976) Über die Lokalisierung von Hohlräumen in Gestein mit der Methode des vertikalen Schweregradienten. *Neue Bergbautechnik* **6**, 194–7.

33. FAJKLEWICZ, Z. J. (1979) Stand der mikrogravimetrischen Untersuchungen bei der Erkundung von Gebirgshohlraumen. Freiberg Forsch.-H., C341. Hrsg; Bergakademie, Freiberg, Leipzig; VEB Deutscher Verlag für Grundstoffindustrie.
34. DRESEN, L., FAJKLEWICZ, Z. J., GÖTZE, H.-J., SOMMER, H. and TE KOOK, J. (1981) Die Ortung oberflächennaker Hohlräume durch die Bestimmung des Vertikalgradienten der Schwere. Glückauf-Forsch.-H. **42**, 84–8.
35. DRESEN, L. et al. (1981) The locating and mapping of abandoned mine shafts at shallow depths by engineering geophysical methods. Preprint Paper No. E-2.7 presented at the 51st Annual International Meeting of the Soc. of Expl. Geophysicists, Los Angeles, California.
36. BUTLER, D. K. (1981) Microgravimetric techniques for geotechnical applications. Misc. Paper GL-80-13, US Army Engineer Waterways Exp. Sta.
37. MERKEL, R. H. (1971) Resistivity analysis for plane-layer half-space model with buried current sources. Geophys. Prosp. **19**, 626–39.
38. MERKEL, R. H. and ALEXANDER, S. S. (1971) Resistivity analysis for models of a sphere in a half-space with buried current electrodes. Geophys. Prosp. **19**, 640–51.
39. DANIELS, J. J. (1978) Interpretation of buried electrode resistivity data using a layered earth model. Geophysics **43**, 988–1001.
40. DANIELS, J. J. and SCOTT, J. H. (1981) Hole-to-surface resistivity measurements. US Geol. Survey Open File Rep. No. 81-1336.
41. BRIEDEN, J., LÖSCH, W., MILITZER, H. and RÖSLER, R. (1978) Die Anwendung von Geoelektrischen Messystemen mit versenkten Elektroden zur Ortung von Hohlräumen. Neue Bergbautechnik **8**, 191–7.
42. BRIEDEN, H., MILITZER, H. and RÖSLER, R. (1979) Geoelektrischen Zwischenfelderkundungeine efektive Methods zum Hohlraumenachweis. Neue Bergbautechnik **9**(5), 243–7.
43. MILITZER, H., RÖSLER, R. and BRIEDEN, H.-J. (1979) Modellkurven zum Geoelektrischen Hohlraumnachweis—Electrodensysteme in Bohrlöchen. Wissenschaftlichen Informationszentrums der Bergakademie, Freiberg, DDR.
44. PARASNIS, D. S. (1964) Long horizontal cylindrical ore body at arbitrary depth in the field of two linear current electrodes. Geophys. Prosp. **12**, 457–87.
45. LYTLE, R. J. et al. (1976) Using cross-borehole electromagnetic probing to locate a tunnel. Lawrence Livermore Laboratory Rep. UCRL-52166.
46. LYTLE, R. J. et al. (1977) Analysis of electromagnetic wave probing for underground voids. Lawrence Livermore Laboratory Rep. UCRL-52214.
47. LYTLE, R. J. et al. (1979) Cross-borehole electromagnetic probing to locate high-contrast anomalies. Geophysics **44**, 1667–77.
48. OWEN, T. E. and SUHLER, S. A. (1980) Subsurface void detection using surface resistivity and borehole electromagnetic techniques. Preprint Paper No. E-7 presented at the 50th Annual International Meeting of the Soc. of Expl. Geophysicists, Houston, Texas.
49. OWEN, T. E. (1981) Cavity detection using VHF hole-to-hole electromagnetic techniques. Proc. Symposium on Tunnel Detection, Colorado School of Mines, Golden, Colorado (21–23 July), sponsored by US Army Mobility Equipment R&D Command, Fort Belvoir, Virginia.

50. RICHMOND, J. H. (1966) TE-wave scattering by a dielectric cylinder of arbitrary cross-section shape. *IEEE Trans. Ant. Prop.* **AP-14**, 460–4.
51. DRESEN, L. (1973) Investigation of diffracted seismic wave amplitudes as a method for locating circular cylindrical cavities in solid rock. *Proc. Symp. on Sink-Holes and Subsidence Engineering Geological Problems*, Int'd. Assoc. of Eng. Geol., Hannover, FRG.
52. LANGE, H. (1980) Seismische Durchschallungen zum Nachweis von Inhomogenitäten des Gebirges. *Neue Bergbautechnik* **10**(11), 644–8.
53. DRESEN, L. *et al.* (1975) Ortung lines verdeckten Schachtes mit geophysikalischen Methoden. *Glückauf-Forsch.-H.* **36**(5), 209–15.
54. DRESEN, L. and ULLRICH, G. (1976) Modellseismische Üntersuchungen über Querschnitte und gebräcer Zonen bei der Ortung verlassener Schächte. *Glückauf-Forsch.-H.* **37**(3), 81–5.
55. DRESEN, L., CASTEN, U. and ULLRICH, G. (1976) Ingenieurgeophysikalischen Nachweis verlassener Schachte und dressen Uberprufung durch Bohrungen. *Glückauf-Forsch.-H.* **122**(23), 1319–24.
56. LAFEHR, R. T., LAURIN, P. J. and BLACK, A. (1981) Tunnel detection by surface and borehole gravity methods. *Proc. Symposium on Tunnel Detection*, Colorado School of Mines, Golden, Colorado (21–23 July), sponsored by US Army Mobility Equipment R & D Command, Fort Belvoir, Virginia.
57. DINES, K. A. and LYTLE, R. J. (1979) Computerized geophysical tomography. *Proc. IEEE* **67**, 1065–73.

INDEX

Acoustic logs, 79–82
Acoustic probes, 79
Adaptive lag-sum (ALS), 16
Afmag technique, 112
Algebraic reconstruction (ART), 27
Amplitude spectrum, 94, 95
Analog filters, 92
Analog ratemeter, 92–5
Anisotropy effects, 25–6, 116
Anomalous zone definition, 37
Anomaly Effect, 177, 178, 179
Apparent resistivity, 112, 113, 137, 166, 177
   anomalies, 189, 191, 229
   profiles, 197, 224–8, 230, 231, 232, 234
   response, 188, 190
Athabasca sandstone, Saskatchewan, 127–34
Attenuation
   coefficient, 25
   measurement in coal seams, 25
Audiofrequency magnetotelluric (AMT) sounding, 107–59
   applications, 155
   artificial sources, 109, 149
   background to, 108–9
   controlled source, 109, 149–55
   data interpretation, 112–16
   depth of sounding, 110
   field studies, 116
   instrumentation, 116–17

Audiofrequency magnetotelluric (AMT) sounding—*contd.*
   layer effects, 114
   natural source fields, 109
   one-dimensional examples, 117–42
   plane wave sources, 155
   principle of method, 109–17
   results, 117–55
   source fields, 112
   standard curve, 112
   two-dimensional examples, 142–55

Basalt, Colorado, 200
Bessel function, 98
Bipole–bipole array, 230
Bleiberg lead–zinc mine, 47
Bonneville salt flats, 151
Borehole
   electromagnetic techniques, 233–40
   gravity methods, 251–3
   methods, 163
   resistivity techniques, 221–33
   seismic techniques, 240–51
Bouguer
   anomaly maps, 47
   correction, 45
   gravity map, 57, 61

Cavendish sulphide deposits, 149, 152
Caves. *See* Cavity detection and mapping

260 INDEX

Cavity detection and mapping, 161–258
 applications, 161–3
 borehole detection methods, 221–53
 characteristic resistivity response,
   185–91
 data-processing techniques, 191–204
 electrical resistivity methods, 164–204
 geophysical techniques, 162, 163
 reconnaissance surveys, 162
 surface gravity methods, 208–21
 surface seismic methods, 204–8
 targets of interest, 162
Cavity Effect, 177, 178
CDP stacking, 15, 16, 17
Chalk River, 142–9
Channel wave(s), 6–9
geotomography, 26–8
Coal
 exploration methods, 4
 mining
  deep, 1
  geophysics of, 2
 seams
  attenuation in, 25
  fault detection, 1–34
  seismic waves in, 6
Computer
 programs, 75
 simulation, 83
Convolution
 integral, 66
 operation, 66
Cosine bell, 77, 78

Dakto sandstone, 136, 137, 141
Deconvolution
 effect, 68, 76, 84, 89, 98
 filter, 87–9
Density
 determination, 36, 43, 62
 profiles, 42, 44
Differentiation filter, 87–9
Diffraction theory, Kirchhoff–Huygens
   approach to, 15
Digital convolution, 67, 68
Digital filtering, 65–105
 exact digital filters, 72–3

Digital filtering—*contd.*
 noise rejection, 75–9
 optimum filters, 74–5
 principles of, 66
 see also under specific applications
Digital logging systems, 65
Digital time series analysis techniques,
   66–79
Discrete convolution, 67
Dispersion estimation, 11
Dynamic trace gathering (DTG), 16, 17,
   18, 26

Earth resistivity
 data processing program, 198
 electrode arrays, 166–71
 field, 199
 measurements, 164
 surveys, 179, 200
Earth Resistivity Data Acquisition
   System (ERDAS), 179–85, 195
Eismänner Kluft, 62
Electrical resistivity methods, 164–204
Electromagnetic impedance, 109
Exponential analog ratemeter, 78

Focused electrode array, 232
Fourier
 analysis, 66
 transform, 7, 11, 12, 13, 25, 72, 86,
   98, 103
Frequency
 domain methods, 72
 spectrum, 103
Frequency–wavenumber surface, 26

Gamma-ray logs, 76, 77, 95–101, 103
Gibb's phenomenon, 77
Gold Hill, Colorado, 236
Gravimeter, 36
Gravity
 anomalies, 219
 anomaly map, 52, 54
 effect profiles, 210–13

Gravity—contd.
  gradient
    field survey results, 220
    instrumentation, 218
    measurements, 217, 219
    vertical, 214, 215, 216, 219
  profiles, 51, 55, 61
  prospecting method, 35
  trend map, 53
Green's function, 7

Herren-Kunst Mine, Eschweiler, FRG, 250
Herren-Kunst Mine Shaft, 247
Hole-to-hole
  electromagnetic tests, 241–3
  electromagnetic wave transmission techniques, 233–40
  seismic techniques, 245
Hole-to-surface tests, 247–50

Impedance measurement, 112, 116
Impulse response, 66–7, 78, 92, 93
In-seam seismology, 1–34
  alternative sources, 28–9
  anisotropy effects, 25–6
  applications, 24–9
  confidence levels, 24
  data acquisition, 9–10
  detection capabilities, 20–1
  dispersion estimation, 11
  early history of, 13
  examples of, 17–24
  principle of method, 5–6
  problems and progress, 25
  processing of reflection surveys, 15–24
  reflection surveys, 10
  research, 6
  routine aid, as, 30
  transmission surveys, 10
  verification, 24
Inverse filter, 80, 86–7, 93, 97, 98, 99

Kirchhoff–Huygens approach to diffraction theory, 15

Lacoste and Romberg borehole gravity meter, 251
Lag-sum, 15
Least-square error (LSE) filters, 74–5, 84
Longwall faces, 3
Low-pass filter, 77, 78

Magnetotelluric sounding. See Audiofrequency magnetotelluric (AMT) sounding
Maxwell's equations, 116
Medford Cave, Florida, 192, 194, 196, 239–43
Mesozoic/Palaeozoic sediments, New Mexico, 134–41
Mine workings, 162, 199, 202, 203, 208, 246, 247

Noise rejection, 75–9

Oil crisis, 1
Optimum filters, 74

Padé approximant, 11
Palaeozoic sedimentary basin of southern Ontario, 117–20
Permafrost mapping, 141–2
Phase
  correction, 17
  roll factor, 17
Pole–dipole
  electrode array, 164, 172
  resistivity surveys, 229
  response characteristics, 189
  survey technique, 173–5
Porosity determination, 36
Potential measurement electrode array, 184

Q value, 25

Radioactive decay processes, 77
Radioelement
  concentrations, 101
  distribution, 100
Rayleigh waves, 204–8, 244
Reflection tomography, 27
Refraction methods, 204
Resistivities, near-surface, 111
Resistivity
  anomalies, 202, 203
  frequency plot, 112
  logs, 82–5
  mapping, 155
  profile, 88
  response zone, 172
Response equalisation filtering, 101–3

Schlumberger
  depth-sounding profile, 193
  type array, 173
Schwarz–Christoffel transformation, 197
Seismology, in-seam. *See* In-seam seismology
Shear waves, 204
Sinc function, 77, 78
Skin
  depth, 110–11, 117, 155
  effect, 84
Smoothing
  effect, 68, 76, 92
  filter, 96
Southern Ontario, Palaeozoic sediments, 117–20
Stefani shaft, 59, 60, 61
Subsidence problems, 162
Sulphide exploration, 149, 152
Surface
  boreholes, 4
  electrode arrays, 174
  geophysics, 5
  gravity surveys, 208–21
  seismic exploration techniques, 204–8
System response function, 97, 98

Taylor expansion, 13, 29
Telemetry systems, 10
Temperature logs, 85–9
Thermal gradient logs, 89
Thermal resistivities, 91
Time-domain
  function, 77
  inverse filter, 86
Time-series analysis, 66
Trace envelope stacking, 17
Tunnels
  air-filled, 210, 214, 216, 252, 253
  water-filled, 210, 214, 216
  see also Cavity detection and mapping

Underground cavities. *See* Cavity; Tunnels
Underground gravity surveys, 35–63
  applications, 41–5
  data presentation, 47
  examples, 47–62
  horizontal, 45–62
  vertical, 36, 38–45
Union Kluft, 55, 59

VHF radar techniques, 163

Wenner array, 165, 172, 174
Wetter-on-the-Ruhr, FRG, 249, 250, 251
Wiener filter, 74
Williston Basin, Manitoba and Saskatchewan, 120–7

Z-transform, 73, 93
Zuni area of New Mexico, 134–41